Student Study Guide

for use with

Calculus
Concepts and Connections
First Edition

Robert T. Smith
Millersville University

Roland B. Minton
Roanoke College

Prepared by

Martina Bode
Northwestern University
and
Donna S. Krichiver
Johnson County Community College

Boston   Burr Ridge, IL   Dubuque, IA   Madison, WI   New York   San Francisco   St. Louis
Bangkok   Bogotá   Caracas   Kuala Lumpur   Lisbon   London   Madrid   Mexico City
Milan   Montreal   New Delhi   Santiago   Seoul   Singapore   Sydney   Taipei   Toronto

Dedicated to our students.

The McGraw·Hill Companies

Student Study Guide for use with
CALCULUS: CONCEPTS AND CONNECTIONS, FIRST EDITION
ROBERT T. SMITH AND ROLAND B. MINTON

Published by McGraw-Hill Higher Education, an imprint of The McGraw-Hill Companies, Inc., 1221 Avenue of the Americas, New York, NY 10020. Copyright © 2006 by The McGraw-Hill Companies, Inc. All rights reserved.

No part of this publication may be reproduced or distributed in any form or by any means, or stored in a database or retrieval system, without the prior written consent of The McGraw-Hill Companies, Inc., including, but not limited to, network or other electronic storage or transmission, or broadcast for distance learning.

This book is printed on acid-free paper.

1 2 3 4 5 6 7 8 9 0 QPD/QPD 0 9 8 7 6 5

ISBN 0-07-303041-4

www.mhhe.com

Contents

Preface ... v

Chapter 0. Preliminaries ... 1
- 0.1. Polynomials and Rational Functions ... 1
- 0.2. Graphing Calculators and CAS ... 8
- 0.3. Inverse Functions ... 13
- 0.4. Trigonometric and Inverse Trigonometric Functions ... 16
- 0.5. Exponential and Logarithmic Functions ... 20
- 0.6. Transformations of Functions ... 25
- 0.7. Parametric Equations and Polar Coordinates ... 31

Chapter 1. Limits and Continuity ... 38
- 1.1. A Brief Preview of Calculus ... 38
- 1.2. The Concept of Limit ... 43
- 1.3. Computation of Limits ... 51
- 1.4. Continuity And Its Consequences ... 60
- 1.5. Limits Involving Infinity ... 66
- 1.6. Limits and Loss-Of-Significance Errors ... 74

Chapter 2. Differentiation ... 77
- 2.1. Tangent Lines and Velocity ... 77
- 2.2. The Derivative ... 86
- 2.3. The Power Rule ... 97
- 2.4. Product and Quotient Rule ... 102
- 2.5. The Chain Rule ... 107
- 2.6. Derivatives of Trigonometric Functions ... 111
- 2.7. Derivatives of Exponential and Logarithmic Functions ... 117
- 2.8. Implicit Differentiation and Inverse Trigonometric Functions ... 124
- 2.9. The Mean Value Theorem ... 133

Chapter 3. Applications of Differentiation — 139

- 3.1. Linear Approximation and Newton's Method — 139
- 3.2. Indeterminate Forms and L'Hôpital's Rule — 147
- 3.3. Maximum and Minimum Values — 156
- 3.4. Increasing And Decreasing Functions — 162
- 3.5. Concavity And Overview Of Curve Sketching — 169
- 3.6. Optimization — 178
- 3.7. Rates of Change in Economics and the Sciences — 187
- 3.8. Related Rates and Parametric Equations — 191

Chapter 4. Integration — 199

- 4.1. Area Under A Curve — 199
- 4.2. The Definite Integral — 206
- 4.3. Antiderivatives — 217
- 4.4. The Fundamental Theorem of Calculus — 226
- 4.5. Integration by Substitution — 230
- 4.6. Integration By Parts — 238
- 4.7. Other Techniques of Integration — 246
- 4.8. Integration Tables — 257
- 4.9. Numerical Integration — 260
- 4.10. Improper Integrals — 265

Chapter 5. Applications of the Definite Integral — 275

- 5.1. Area of a Plane Region — 275
- 5.2. Volume — 281
- 5.3. Arc Length and Surface Area — 288
- 5.4. Projectile Motion — 293
- 5.5. Applications of Integration to Physics and Engineering — 299
- 5.6. Probability — 306

Chapter 6. Differential Equations — 312

- 6.1. Growth and Decay Problems — 312
- 6.2. Separable Differential Equations — 319
- 6.3. Direction Fields and Euler's Method — 325
- 6.4. Second Order Equations with Constant Coefficients — 330
- 6.5. Nonhomogeneous Equations: Undetermined Coefficients — 333

Chapter 7. Infinite Series — 335

- 7.1. Sequences of Real Numbers — 335
- 7.2. Infinite Series — 341
- 7.3. The Integral Test and Comparison Tests — 346
- 7.4. Alternating Series — 353
- 7.5. Absolute Convergence And The Ratio Test — 357
- 7.6. Power Series — 363
- 7.7. Taylor Series — 369
- 7.8. Applications of Taylor Series — 375
- 7.9. Fourier Series — 378
- 7.10. Using Series to Solve Differential Equations — 383

Preface

How to use this Study Guide. This Study Guide is designed to be used parallel to the text. Each section has a collection of worked out problems focusing on the key topics. This guide can be used before the material is covered in class, or after it has been covered. You should gain a more solid understanding and grasp of the key topics in each section by working through these examples. Always try to work through the examples first by covering up the solution, then double check your solution by comparing it to the solution in the guide.

Organization. The sections in this guide correspond to the sections in the text. Each section begins with an explanation of the key topics, followed by a collection of worked out example problems. After stating the example problems, the solution starts with a strategy part. When working through these problems first cover up the strategy, solution and summary parts, and see how far you can get working out the problem by yourself. If you get stuck, uncover the strategy part, and see if you can now solve the problem. If you need to see the entire solution uncover the solution. Each problem has a summary part that shortly summarizes the difficult and interesting points of each problem. Remember to always try to work the problems on your own before uncovering the solution.

CHAPTER 0

Preliminaries

0.1. Polynomials and Rational Functions

Key Topics.
- Equations of Lines
- Functions
- Quadratic Formula

Worked Examples.
- Finding the Equation of a Line Given Two Points
- Evaluating a Function
- Finding the Domain of a Function
- Solving an Equation using the Quadratic Formula

Overview

This chapter is a review of precalculus. We are not going to write an entire precalculus book here, but we are going to focus on the topics that are fundamental to the study of calculus. There will probably be some topics that you do not remember as well as others. In that case, use this study guide, and work through more examples in the textbook. If you need more help, go to a precalculus, college algebra, or trigonometry text where the topics are covered in greater depth.

Equations of Lines. In algebra a lot of time is spent on learning how to find line equations. In calculus this knowledge is applied to fundamental calculus concepts. Finding slopes and line equations are key elements of calculus.

Functions. We will work with functions throughout all of calculus. A function is a rule that assigns exactly one element y, a number in the range, to an element x, a number in the domain.

Quadratic Formula. A quadratic equation is a second degree equation. All quadratic equations can be solved using the quadratic formula. The solutions to the equation
$$ax^2 + bx + c = 0$$
are given by
$$x = \frac{-b \pm \sqrt{b^2 - 4ac}}{2a}$$

Example 1: Finding the Equation of a Line Given Two Points

Find an equation of the line through the points $(-2, 3)$ and $(-7, -4)$.

Strategy. We will find the slope using the formula for slopes:
$$m = \frac{y_2 - y_1}{x_2 - x_1}$$
Then we use the point-slope formula with one of the given points:
$$y - y_1 = m(x - x_1)$$

Solution.
$$m = \frac{-4 - 3}{-7 - (-2)} = \frac{-7}{-7 + 2} = \frac{-7}{-5} = \frac{7}{5}$$

We can use the point $(-2, 3)$, or the point $(-7, -4)$ with the point-slope equation to find the line equation. Using $(-2, 3)$, we have

$$\begin{aligned}
y - 3 &= \frac{7}{5}(x - (-2)) \\
y - 3 &= \frac{7}{5}(x + 2) \\
y &= \frac{7}{5}x + \frac{14}{5} + 3 \\
y &= \frac{7}{5}x + \frac{29}{5}
\end{aligned}$$

Summary. In order to find the equation of a line, we need the slope of the line and one point. Make sure you understand how to do this, since finding slopes and writing the equations of lines is fundamental to calculus.

Example 2: Evaluating a Function

Given $f(x) = 3x^2 - x + 7$, find $f(2)$, $f(-1)$, $f(a)$, $f(a+h)$, and $\dfrac{f(a+h) - f(a)}{h}$.

Strategy. In each case, we need to go to the function $f(x)$ and replace the x with the given number or quantity.

Solution. a) $f(2) = 3(2)^2 - (2) + 7 = 17$
b) $f(-1) = 3(-1)^2 - (-1) + 7 = 3 + 1 + 7 = 11$
c) $f(a) = 3(a)^2 - (a) + 7 = 3a^2 - a + 7$
d)
$$\begin{aligned} f(a+h) &= 3(a+h)^2 - (a+h) + 7 \\ &= 3(a^2 + 2ah + h^2) - a - h + 7 \\ &= 3a^2 + 6ah + 3h^2 - a - h + 7 \end{aligned}$$

e)
$$\begin{aligned} \dfrac{f(a+h) - f(a)}{h} &= \dfrac{(3a^2 + 6ah + 3h^2 - a - h + 7) - (3a^2 - a + 7)}{h} \\ &= \dfrac{3a^2 + 6ah + 3h^2 - a - h + 7 - 3a^2 + a - 7}{h} \\ &= \dfrac{6ah + 3h^2 - h}{h} \quad \text{(factor out } h\text{)} \\ &= \dfrac{h(6a + 3h - 1)}{h} \\ &= 6a + 3h - 1 \end{aligned}$$

Summary. When you evaluate a function take the x out and put in its replacement. It is very important that you put parentheses around the replacement, especially if there is more than one symbol. For example to replace x with -7, in $3x^2$, we need $3(-7)^2 = 3(49) = 147$.

Example 3: Finding the Domain of a Function

Find the domain of

$$\text{a) } f(x) = \sqrt{2x+7}$$
$$\text{b) } g(x) = \frac{2x-1}{5x+4}$$

Strategy. In order to find the domain, we start by asking what must happen for this function to be defined. The expression under a square root must be non-negative. The denominator of a rational function cannot be zero.

Solution. a) $f(x) = \sqrt{2x+7}$

Since we are taking the square root of $2x+7$, that expression cannot be negative. We need to solve the inequality

$$\begin{aligned} 2x+7 &\geq 0 \\ 2x &\geq -7 \\ x &\geq -\frac{7}{2} \end{aligned}$$

The domain of the function $f(x) = \sqrt{2x+7}$ is $[-\frac{7}{2}, \infty)$.

b) $g(x) = \dfrac{2x-1}{5x+4}$

The rational function is defined for all real numbers x, except for those values which make the denominator equal to zero. In this problem, we solve

$$\begin{aligned} 5x+4 &= 0 \\ 5x &= -4 \\ x &= -\frac{4}{5} \end{aligned}$$

We can express the domain by simply writing $x \neq -\frac{4}{5}$, or by writing the domain in interval notation as $(-\infty, -\frac{4}{5}) \cup (-\frac{4}{5}, \infty)$.

Summary. In part a) we looked for values of x which would keep the radicand (the expression under the radical) non-negative.

In part b) we had $x \neq -\frac{4}{5}$, which tells us that we can choose any number for x except for $-\frac{4}{5}$.

Example 4: Solving an Equation using the Quadratic Formula

Solve the following quadratic equations using the quadratic formula.
a) $2x^2 - 35 = -9x$
b) $x^2 + 6x + 7 = 0$
c) $x^2 - 4x = -13$

Strategy. In all cases, we arrange the terms in such a way that one side is equal to zero. Then we substitute the constants into the quadratic formula.

$$ax^2 + bx + c = 0 \implies x = \frac{-b \pm \sqrt{b^2 - 4ac}}{2a}$$

Solution. a) $2x^2 - 35 = -9x \implies 2x^2 + 9x - 35 = 0$
Thus $a = 2$, $b = 9$, and $c = -35$, and we have

$$\begin{aligned} x &= \frac{-9 \pm \sqrt{9^2 - 4(2)(-35)}}{2(2)} \\ &= \frac{-9 \pm \sqrt{81 + 280}}{4} \\ &= \frac{-9 \pm \sqrt{361}}{4} \\ &= \frac{-9 \pm 19}{4} \end{aligned}$$

Therefore the two solutions are

$$x_1 = \frac{-9 + 19}{4} = \frac{10}{4} = \frac{5}{2}$$

and

$$x_2 = \frac{-9 - 19}{4} = \frac{-28}{4} = -7$$

b) $x^2 + 6x + 7 = 0$
In this example, we have $a = 1$, $b = 6$, and $c = 7$.

$$\begin{aligned} x &= \frac{-6 \pm \sqrt{6^2 - 4(1)(7)}}{2(1)} \\ &= \frac{-6 \pm \sqrt{8}}{2} = \frac{-6 \pm 2\sqrt{2}}{2} = \frac{2(-3 \pm \sqrt{2})}{2} = -3 \pm \sqrt{2} \end{aligned}$$

The two solutions are

$$x_1 = -3 + \sqrt{2}, \text{ and } x_2 = -3 - \sqrt{2}$$

c) $x^2 - 4x = -13 \implies x^2 - 4x + 13 = 0$
Here we have $a = 1$, $b = -4$, and $c = 13$

$$x = \frac{4 \pm \sqrt{(-4)^2 - 4(1)(13)}}{2(1)} = \frac{4 \pm \sqrt{-36}}{2}$$

Since $\sqrt{-36}$ is not a real number, this quadratic equation has no real solutions. Remembering that $\sqrt{-1} = i$, we can find the two imaginary solutions:
$$x = \frac{4 \pm 6i}{2} = \frac{2(2 \pm 3i)}{2} = 2 \pm 3i$$
The two solutions are:
$$x_1 = 2 + 3i, \text{ and } x_2 = 2 - 3i$$

Summary. Make sure to arrange the terms in the right order before applying the quadratic formula. Then identify the constants a, b, and c and plug the numbers into the quadratic equation.

0.2. Graphing Calculators and CAS

Key Topics.
- Local Extrema
- Inflection Points
- Vertical Asymptotes

Worked Examples.
- Finding Extrema from a Graph
- Finding Inflection Points from a Graph
- Finding Vertical Asymptotes of a Rational Function

Overview

In this section we will give special names to specific points that we see on a graph. As the course proceeds, we will learn methods of calculus to exactly identify these *special* points. Instead of approximating what they are by looking at a graph, we will have ways to analytically find them.

Local Extrema. Extrema is the plural of the word extreme. The extremes or extrema of a function are the highest and the lowest points of the function. Often, these are $\pm\infty$. We really do not need to know that we can make the maximum amount of money by selling an infinite number of T-shirts, right? Instead we look at the highest (or lowest) point of a function in a certain locality - hence, the name *local extrema*. Many books also call these relative extrema since they are the maximum or minimum points relative to the interval surrounding them.

Inflection Points. We are not only interested in where a graph is increasing or decreasing, but whether it curves upwards (concave up) or curves downward (concave down) while it is increasing. Look at, the graph of $y = x^3$.

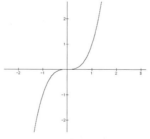

The graph is always increasing, but it is curving downwards on the interval $(-\infty, 0)$ and curving upwards on the interval $(0, \infty)$. The point where the concavity changes $(0,0)$ - where it switches from concave up to concave down, or vice versa, is called an inflection point. We will see many inflection points in later chapters.

Vertical Asymptotes. Vertical asymptotes are of the form $x = a$, and occur at values of x which are not in the domain. A very common place for vertical asymptotes is at the values of x in a rational function, which cause the denominator to equal zero. Other examples of functions that have vertical asymptotes are $y = \ln x$, and $y = \tan x$. The function $y = \ln x$ has a vertical asymptote at $x = 0$, and $y = \tan x$ has vertical asymptotes at $x = -\frac{\pi}{2}$, and $x = \frac{\pi}{2}$.

Example 1: Finding Extrema from a Graph

Find all extrema, and intercepts of the function $f(x) = \frac{1}{2}x^4 - x^2$. The graph of f is shown below.

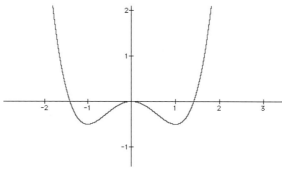

Strategy. The extrema occur at the highest and lowest points of the graph. Intercepts occur where the graph crosses the x-axis, or the y-axis.

Solution. From the graph, we can see that f has local minima at $x = -1$, and $x = 1$, and a local maximum at $x = 0$.

f is decreasing for $x < -1$, and $0 < x < 1$, and f is increasing for $-1 < x < 0$, and $x > 1$.

The y-intercept is equal to $f(0) = 0$. The x-intercepts are solutions of

$$\frac{1}{2}x^4 - x^2 = 0$$
$$x^2\left(\frac{1}{2}x^2 - 1\right) = 0$$

Thus $x = 0$, or $\frac{1}{2}x^2 - 1 = 0$.

$$\frac{1}{2}x^2 - 1 = 0 \text{ for } x^2 = 2, \text{ thus } x = \pm\sqrt{2}$$

$$\implies x\text{-intercepts} = (0,0) \text{ and } (\pm\sqrt{2}, 0)$$

Summary. If we are looking for y-intercepts, we are looking for points on the y-axis, therefore points where $x = 0$.

Similarly, when we are looking for x-intercepts, we are looking for points on the x-axis, therefore points where $y = 0$.

Example 2: Finding Inflection Points from a Graph

Find all inflection points of the function $f(x) = \sin x$ for $-\pi \leq x \leq 2\pi$.

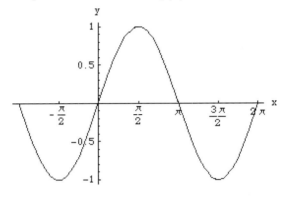

Strategy. Note that the graph is concave up, switching to concave down, and then switching to concave up again. The points where the graph changes concavity are the inflection points.

Solution. Another way of thinking of the concavity is:
f bends upward for $-\pi \leq x \leq 0$, and for $\pi \leq x \leq 2\pi$.
f bends downward for $0 \leq x \leq \pi$.
Therefore the inflection points are $(0, 0)$ and $(\pi, 0)$.

Summary. Be aware of the fact that 0 is at the end of the interval where the graph *bends upward*, and also 0 is at the beginning of the interval where the graph *bends downward*. That is how we know that it is the x-coordinate of the point of inflection. In a similar manner, we find any other points of inflection.

Example 3: Finding Vertical Asymptotes of a Rational Function

---

Find all vertical asymptotes of the function
$$f(x) = \frac{2x - 2}{x^2 - 1}$$

Strategy. This is a rational function. The denominator cannot be equal to zero. There will be a vertical asymptote at any point where the denominator is zero **after we have reduced the expression.**

Solution. Possible asymptotes are located at $x = -1$ and $x = 1$. However, there is no vertical asymptote at $x = 1$. Algebraically, this is because we can first reduce the function
$$f(x) = \frac{2x - 2}{x^2 - 1} = \frac{2(x - 1)}{(x - 1)(x + 1)} = \frac{2}{x + 1}$$
Therefore, the only vertical asymptote occurs at $x = -1$ as shown in the graph.

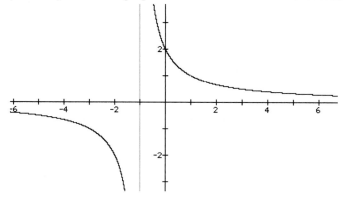

Summary. The point $x = 1$ which also causes the denominator to equal zero is a point of discontinuity, meaning that the function is not continuous there, but there is no vertical asymptote. We call that point a removable discontinuity.

0.3. Inverse Functions

Key Topics.
- Definition of Inverse Functions
- Notation of Inverse Functions
- Graph of Inverse Functions

Worked Examples.
- Finding the Inverse of a Function
- Graphing a Function and the Inverse of the Function

Overview

In this section we continue our study of functions, by working with *inverse* functions. The word *inverse* is used in a similar manner as it is in everyday English usage. The *inverse* of sitting down is standing up. The *inverse* of putting a book down, is picking the book up, etc.

Definition of Inverse Functions. Often a function has an inverse. The function and its inverse *undo* each other. A simple example is adding and subtracting. For example, if we start with a number and add 5 to it and then subtract 5 from the answer, we are back to the original number. In mathematical terms, if $f(x)$ and $g(x)$ are inverse functions, then $f(g(x)) = g(f(x)) = x$.

Notation of Inverse Functions. If $f(x)$ is a function and $g(x)$ is its inverse, then we write $g(x) = f^{-1}(x)$. Read as *f inverse of x*. Note that $f^{-1}(x) \neq \dfrac{1}{f(x)}$, although it certainly looks like it should. Remember that the exponent -1 on a function refers to the inverse of that function.

Graph of Inverse Functions. When two functions are inverses of each other, their graphs are symmetric to the line $y = x$.

Example 1: Finding the Inverse of a Function

If $f(x) = \sqrt{2x^3 + 1}$, then find $f^{-1}(x)$.

Strategy. The following steps are used to find the inverse of a function:
1. Replace $f(x)$ with y.
2. Switch the variables x and y.
3. Solve for y.
4. Replace y with $f^{-1}(x)$.

Solution.
$$\begin{aligned} f(x) &= \sqrt{2x^3 + 1} \\ y &= \sqrt{2x^3 + 1} \quad \text{for} \quad y \geq 0 \quad &&(\text{switch } x \text{ and } y) \\ x &= \sqrt{2y^3 + 1} \quad \text{for} \quad x \geq 0 \\ x^2 &= 2y^3 + 1 \\ \frac{x^2 - 1}{2} &= y^3 \\ y &= \sqrt[3]{\frac{x^2 - 1}{2}} \\ f^{-1}(x) &= \sqrt[3]{\frac{x^2 - 1}{2}} \end{aligned}$$

Summary. When finding the inverse of a function, the x and y change places. Therefore, the range of one function is the domain of the other. In the above problem, the range of $f(x)$ was $y \geq 0$. Notice that the domain of $f^{-1}(x)$ is $x \geq 0$.

Example 2: Graphing a Function and the Inverse of the Function

Graph $f(x) = \sqrt{2x^3 + 1}$ and $f^{-1}(x) = \sqrt[3]{\dfrac{x^2 - 1}{2}}$, $x \geq 0$ on the same axes.

Strategy. Using a graphing calculator or by plotting points, graph both functions and obtain the following graph.

Solution. Graph $y = \sqrt{2x^3 + 1}$. Note that because we have a square root, we have $y \geq 0$. Therefore, when graphing the inverse function we only get x-values which are non-negative. Graph both functions

$$f(x) = \sqrt{2x^3 + 1}$$
$$f^{-1}(x) = \sqrt[3]{\dfrac{x^2 - 1}{2}}; \ x \geq 0$$

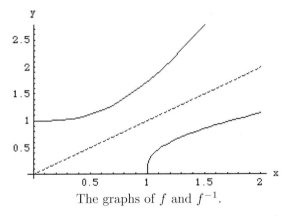

The graphs of f and f^{-1}.

Summary. Once you have both graphs on the same axes, also graph $y = x$. See how the graph of $f(x)$ and the graph of $f^{-1}(x)$ are symmetric to the graph of $y = x$. This is true for all pairs of inverse functions.

0.4. Trigonometric and Inverse Trigonometric Functions

Key Topics.
- Periodic Functions
- Inverse Trigonometric Functions
- Identities

Worked Examples.
- Finding the Period of a Function
- Evaluating Inverse Trigonometric Functions

Overview

We now extend our study of functions to the trigonometric functions. You will use a lot of trigonometry in calculus and will have a greater appreciation of its use by the end of the course.

Periodic Functions. A function is periodic if its basic shape is repeated over and over again at regular intervals. In mathematical symbols, if a function has period T, then $f(x+T) = f(x)$. The trigonometric (trig) functions are the most common periodic functions.

Inverse Trigonometric Functions. When we find the trig function of an angle, we get a ratio. The inverse function, then, should take the ratio and give us back our angle. The notation for inverse trig functions is, appropriately, $\sin^{-1}(x)$, $\cos^{-1}(x)$, $\tan^{-1}(x)$ etc. On your calculator, the buttons which have these notations are the second functions of the same button with $\sin x$, $\cos x$, and $\tan x$, respectively. The inverse trig buttons are *angle getting* buttons. We enter the ratio (usually in the form of a decimal) and the calculator returns the angle.

Because these functions are periodic, there are, for example, an infinite number of angles which would correctly answer the question: What angle has a sine of .5? Some answers to that question are $\frac{\pi}{6}, \frac{5\pi}{6}, \frac{13\pi}{6}, \frac{17\pi}{6}$ etc. But a function only has one answer and we called this an inverse **function**. Consequently, we have to restrict the domain of $\sin y = x$. $y = \sin x$ is restricted to the interval $\left[\frac{-\pi}{2}, \frac{\pi}{2}\right]$. In other words we restrict the sine function to the principle value in quadrants I and IV. See picture.

0.4. TRIGONOMETRIC AND INVERSE TRIGONOMETRIC FUNCTIONS

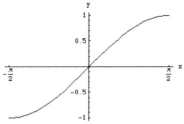

$y = \sin x$ restricted to the domain $[-\frac{\pi}{2}, \frac{\pi}{2}]$

Similarly, we restrict $\tan y = x$ in exactly the same way. For $\cos y = x$, we restrict y to the principle value in quadrants I and II and, hence, to the interval $[0, \pi]$.

Identities. There are a few, not a lot of, identities, which you absolutely must know in order to study calculus. These include reciprocal identities and the Pythagorean identities. You need to know the following:

1. $\csc \theta = \dfrac{1}{\sin \theta}$
2. $\sec \theta = \dfrac{1}{\cos \theta}$
3. $\tan \theta = \dfrac{\sin \theta}{\cos \theta}$
4. $\cot \theta = \dfrac{1}{\tan \theta}$

5. $\sin^2 \theta + \cos^2 \theta = 1$
6. $1 + \tan^2 \theta = \sec^2 \theta$
7. $1 + \cot^2 \theta = \csc^2 \theta$
8. $\sin 2\theta = 2 \sin \theta \cos \theta$

Example 1: Finding the Period of a Function

Find the period of the following periodic functions.

a) $f(x) = \begin{cases} 1 \text{ if } -\pi < x \leq 0 \\ 2 \text{ if } 0 < x \leq 2\pi \\ 1 \text{ if } \pi < x \leq 2\pi \\ 2 \text{ if } 2\pi < x \leq 3\pi \\ \qquad \text{etc.} \end{cases}$

b) $g(x) = \sin(\pi x)$

Strategy. Notice that the values of the function in part a) repeat. The value is either 1 or 2 and the values appear in a predictable pattern. We need to identify the length of the interval of one entire cycle.

In part b) we can use the general formula, which says that the period of the function $y = \sin(Ax)$ is $\dfrac{2\pi}{A}$.

Solution. a) We see that $f(x) = 1$ for $-\pi < x \leq 0$, and $f(x) = 2$ for $0 < x \leq \pi$. We also note that this represents a repeating pattern, and we see that the length of one interval before the function repeats is 2π. The interval goes from $-\pi$ to π. Therefore, the length of one period of this function is 2π.

b) The period of $g(x) = \sin(\pi x)$ is $\dfrac{2\pi}{\pi} = 2$. We can also see this from the graph below.

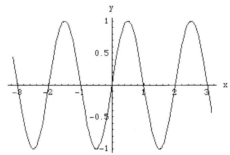

Summary. The function in part a) was not a trig function, because you should recognize that all periodic functions are not trig functions. In reality, however, trig functions are definitely the most familiar periodic functions. If you look at the graph of any of the trig functions, you can identify the length of one period, or cycle, of the graph.

Example 2: Evaluating Inverse Trigonometric Functions

Evaluate a) $y = \tan^{-1} 1$ and b) $y = \sec^{-1} 2$.

Strategy. We rewrite these expressions using the definition of the inverse trigonometric functions. Then we will recognize the *special angle*, which gives the correct answer.

Solution. a) $y = \tan^{-1} 1 \implies \tan y = 1$

We are looking for an angle, y, in quadrants I or IV such that its tangent is equal to 1.

We know that $\tan \dfrac{\pi}{4} = 1$. Therefore, $y = \dfrac{\pi}{4}$.

b) $y = \sec^{-1} 2 \implies \sec y = 2$

We can also write this as: $\cos y = \dfrac{1}{2}$

We are looking for an angle, y, in quadrants I or II such that its cosine is equal to $\dfrac{1}{2}$. We know that $\cos \dfrac{\pi}{3} = \dfrac{1}{2}$. Therefore, $y = \dfrac{\pi}{3}$.

Summary. Along with the identities you should also know the values of all the trigonometric functions of $0, \dfrac{\pi}{2}, \pi, \dfrac{3\pi}{2}, \dfrac{\pi}{6}, \dfrac{\pi}{3}, \dfrac{\pi}{4}$ and all multiples of these *special* angles.

0.5. Exponential and Logarithmic Functions

Key Topics.
- Rules of Exponents
- Solving Logarithmic Equations
- Solving Exponential Equations

Worked Examples .
- Evaluating Exponential Expressions
- Solving Logarithmic Equations
- Solving Exponential Equations

Overview

Now we extend our study of functions to a special pair of inverse functions, the exponential and logarithmic functions.

Rules of Exponents. Given $x > 0$, we have the following important rules of exponents that you should know and fully understand:

a) $x^a \cdot x^b = x^{a+b}$

b) $\dfrac{x^b}{x^a} = x^{b-a}$

c) $(x^a)^b = x^{ab}$

d) $x^{\frac{a}{b}} = \sqrt[b]{x^a} = (\sqrt[b]{x})^a$

e) $x^{-a} = \dfrac{1}{x^a}$

Solving Logarithmic Equations. To solve a logarithmic equation, we first isolate the logarithm on one side of the equation. Then we use the definition of a logarithm to rewrite the equation in exponential form. The new equation should be easily solvable.

Important rules of logarithms that you should know and fully understand are:

a) $y = \log_b x \iff x = y^b$

b) $\log_b b^x = x$

c) $b^{\log_b x} = x, x > 0$

d) $\ln e^x = x$

e) $e^{\ln x} = x, x > 0$

f) $\log_b (xy) = \log_b x + \log_b y$

g) $\log_b \left(\dfrac{x}{y}\right) = \log_b x - \log_b y$

h) $\log_b (x^y) = y \log_b x$ (*exponents jump over logs*)

Solving Exponential Equations. To solve an exponential equation, we first isolate the exponential on one side of the equation. Then we take the natural logarithm on both sides and use the property that says *exponents jump over logs* to solve the equation. For example, $\ln e^{3x} = 3x \ln e = 3x$ since $\ln e = 1$.

Example 1: Evaluating Exponential Expressions

Evaluate a) $8^{\frac{2}{3}}$, b) $8^{-\frac{2}{3}}$, and c) 3^{-2}.

Strategy. We will rewrite the expression so that it is in a form where we can carry out the calculations.

Solution. a) $8^{\frac{2}{3}} = \left(\sqrt[3]{8}\right)^2 = 2^2 = 4$

b) $8^{-\frac{2}{3}} = \dfrac{1}{\left(\sqrt[3]{8}\right)^2} = \dfrac{1}{2^2} = \dfrac{1}{4}$

c) $3^{-2} = \dfrac{1}{3^2} = \dfrac{1}{9}$

Summary. Note the similarity in parts a) and b). The negative exponent tells us to move the base to the denominator looking exactly the same as it did in the numerator, and to change the sign of the exponent to positive.

Example 2: Solving Logarithmic Equations

Solve a) $3\ln x = 12$ and b) $2\ln x + \ln 2x = 1$.

Strategy. We isolate the $\ln x$ and then rewrite the expression as an exponential. At that point the equation should be easily solved.

Solution. a)
$$\begin{aligned} 3\ln x &= 12 \\ \ln x &= 4 \\ e^4 &= x \\ x &\approx 54.6 \end{aligned}$$

b)
$$\begin{aligned} 2\ln x + \ln 2x &= 1 \\ \ln x^2 + \ln 2x &= 1 \\ \ln(x^2 \cdot 2x) &= 1 \\ \ln(2x^3) &= 1 \\ e^1 &= 2x^3 \\ \frac{1}{2}e &= x^3 \\ \sqrt[3]{\frac{1}{2}e} &= x \\ x &\approx 1.1077 \end{aligned}$$

Summary. For all logarithmic equations, express as a single log, isolate the log, change to an exponential, and solve.

Example 3: Solving Exponential Equations

Solve a) $e^{3x} = 4$ and b) $4e^x - 1 = 19$.

Strategy. We will isolate the exponential and then take ln (natural logarithm) on both sides. The exponent can then *jump over the log*. Since $\ln e = 1$, the solving is then easy.

Solution. a)
$$\begin{aligned} e^{3x} &= 4 \\ \ln e^{3x} &= \ln 4 \\ 3x \ln e &= \ln 4 \\ 3x &= \ln 4 \\ x &= \frac{\ln 4}{3} \\ &\approx 0.4621 \end{aligned}$$

b)
$$\begin{aligned} 4e^x - 1 &= 19 \\ 4e^x &= 20 \\ e^x &= 5 \\ \ln e^x &= \ln 5 \\ x \ln e &= \ln 5 \\ x &= \ln 5 \\ &\approx 1.609 \end{aligned}$$

Summary. Since exponential functions and logarithmic functions are inverses of each other, we solve exponential functions by isolating the exponential, then *undoing* it by taking the natural logarithm on each side of the equation.

0.6. Transformations of Functions

Key Topics.
- Composition of Functions
- Translation of Graphs

Worked Examples .
- Finding the Composition of Functions $(f \circ g)(x)$, $(g \circ f)(x)$
- Translation of a Given Graph
- Reflection of a Given Graph

Overview

In this section we work with functions of functions. We also learn how to transform the graph of a function.

Composition of Functions. The composition of functions $f(x)$ and $g(x)$ is defined as: $(f \circ g)(x) = f(g(x))$ and $(g \circ f)(x) = (g(f(x))$.

Translation of Graphs. On page 64 in the text, there is an excellent summary of the way graphs are affected by different changes to the function. In particular, we will look at vertical and horizontal translations of a given graph. We will also see what changes occur in the function, to cause the graph to reflect over the x-axis.

Example 1: Finding the Composition of Functions

Given that $f(x) = x^2 - 7$ and $g(x) = x + 1$, find both $(f \circ g)(x)$ and $(g \circ f)(x)$.

Strategy. It helps to rewrite the composition of $(f \circ g)(x)$ as $f(g(x))$ and $(g \circ f)(x)$ as $g(f(x))$.

Solution.
$$\begin{aligned} (f \circ g)(x) &= f(g(x)) = f(x+1) \\ &= (x+1)^2 - 7 \\ &= x^2 + 2x + 1 - 7 \\ &= x^2 + 2x - 6 \end{aligned}$$

$$\begin{aligned} (g \circ f)(x) &= g(f(x)) = g(x^2 - 7) \\ &= (x^2 - 7) + 1 \\ &= x^2 - 6 \end{aligned}$$

Summary. In math, we usually work from the inside out and one place we do that is in composition of functions. We evaluate the inner function first and then the outer. For example, using the same functions as above, $(f \circ g)(2) = f(g(2))$. First we find $g(2) = 3$, and then the outer function $f(3) = 3^2 - 7 = 2$.

Example 2: Translation of a Given Graph

Given the graph of $y = f(x)$ in the figure, graph a) $y = f(x) - 2$ and b) $y = f(x - 2)$.

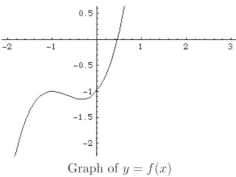

Graph of $y = f(x)$

Strategy. We can identify the points $(-1, -1)$ and $(0, -1)$ on the graph. We can use those points as guides of how to move the graph.

Solution. a) $f(x) - 2$ moves the graph down 2 units. Because we find $f(x)$ first, the -2 affects the y-value and, therefore, moves the graph vertically.

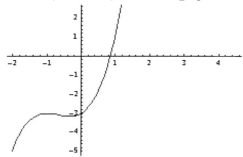

b) $f(x - 2)$ affects the x-value. In other words, we change the x before we apply the function. Therefore, the change is horizontal. The graph moves 2 units to the **right**. Think about this. In the original graph we have $f(0) = -1$. To get the y-value of -1 in the new graph, we need

$$\begin{aligned} f(x - 2) &= f(0) \\ x - 2 &= 0 \\ x &= 2 \end{aligned}$$

We need to be at $x = +2$, which is two units to the right.

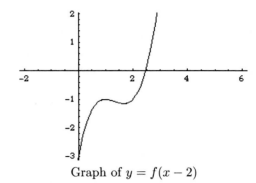

Graph of $y = f(x-2)$

Summary. It is important that you fully understand why the graph moves *backwards* when there is a horizontal shift. Read the explanation above. Try graphing several points and prove it to yourself.

Example 3: Reflection of a Given Graph

Given the graph of $y = \sqrt{x}$, graph each of the following.
a) $y = -\sqrt{x}$
b) $y = -\sqrt{x} + 4$
c) $y = -\sqrt{x+4} + 3$

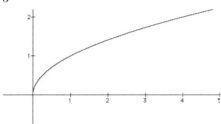

Strategy. We know the shape of the basic graph of $y = \sqrt{x}$. When we look at the changes in the three parts, we first have to decide if y is affected, causing a vertical change, if x is affected causing a horizontal change, or if both x and y are affected.

Solution. a) $y = -\sqrt{x}$

Every value of the original graph will take on the opposite value. For example, in $y = \sqrt{x}$, if $x = 4$, then $y = 2$. In the new graph, $y = -\sqrt{x}$, if $x = 4$, then $y = -2$.

All the values which were above the x-axis switch to going below the x-axis, causing the entire shape to reflect over the x-axis, and we end up with:

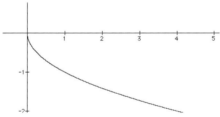

b) $y = -\sqrt{x} + 4$

This graph takes the y-values from part a) and adds 4 to every answer. The result is that the entire shape gets raised up 4 units, nothing else changes - it is like riding up an elevator. The graph looks like:

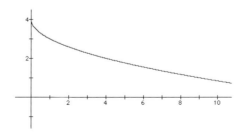

c) $y = -\sqrt{x+4} + 3$

Here several changes occur.

First the expression under the radical goes from x to $x+4$. We have a horizontal change of 4 units to the LEFT.

Secondly, the negative sign causes the graph to reflect over the x-axis.

Lastly, the $+3$ raises the graph up 3 units.

In $y = \sqrt{x}$, the *starting* point is $(0,0)$. In the new graph, the *starting* point is $(-4, 3)$.

Also, the graph now decreases instead of increasing like the original graph, because of the negative sign.

The graph of $y = -\sqrt{x+4} + 3$ looks like:

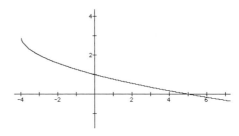

Summary. When there is more than one type of change, make sure that you do each one separately and then put the whole graph together.

0.7. Parametric Equations and Polar Coordinates

Key Topics.
- Parametric Equations
- Polar Coordinates
- Conversion Between Polar and Rectangular Coordinates

Worked Examples .
- A Parametric Graph
- A Polar Graph
- Converting Polar to Rectangular Coordinates
- Converting Rectangular to Polar Coordinates

Overview

Sometimes we need to know the placement of an object in some way other than its horizontal and vertical coordinates. In this section, we learn two other important types of coordinates.

Parametric Equations. Sometimes, instead of $y = f(x)$ we have both x and y as a function of time, t. Not only does it matter where an object is located, it matters when it is located at that point. If you are using a graphing calculator, you need to change the mode to parametric mode. It probably says PAR.

Graphing in parametric mode gives fun graphs from very ordinary looking equations.

Polar Coordinates. Polar coordinates tell us where an object is located by giving the distance and the direction from the origin. The direction is given by an angle. For polar graphs, you have to change the mode to polar. It probably says POL.

Graphing in polar coordinates also results in very interesting graphs.

Conversion Between Polar and Rectangular Coordinates. To move between polar coordinates and rectangular coordinates use the following equations which are found using the Pythagorean Theorem and the definition of the trigonometric functions.

If we know (r, θ) we can get (x, y) by:
$$x = r\cos\theta$$
$$y = r\sin\theta$$

If we know (x, y) we can get (r, θ) by:
$$r^2 = x^2 + y^2$$
$$\tan\theta = \frac{y}{x}$$
$$\theta = \tan^{-1}\left(\frac{y}{x}\right)$$

Example 1: Parametric Graph

Sketch the graph defined by $\begin{cases} x = \cos t \\ y = 2\sin t \end{cases}$.
Find a corresponding rectangular equation for the curve.

Strategy. It is helpful to first create a table of values. The entries of the table are time, t, and the coordinates of the point (x, y) at time t. Then we will use the identity $\cos^2 t + \sin^2 t = 1$ to find the equation in rectangular coordinates.

Solution. We start with a table of values.

t	$x = \cos t$	$y = 2\sin t$
0	1	0
$\frac{\pi}{2}$	0	2
π	-1	0
$\frac{3\pi}{2}$	0	-2
π	1	0

The staring point at time $t = 0$ is the point $(1, 0)$. From there we travel counterclockwise to the point $(0, 2)$ and so on. Alternatively, we can also use a CAS or a graphing calculator to obtain the graph.

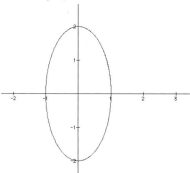

The graph of $x = \cos t$ and $y = 2\sin t$ appears to be an ellipse.

In order to change to rectangular coordinates, we start with $\cos^2 t + \sin^2 t = 1$. We know that:

$$\cos t = x \implies \cos^2 t = x^2$$
$$2\sin t = y \implies \sin^2 t = \frac{y^2}{4}$$

Substituting these values, we then have:

$$x^2 + \frac{y^2}{4} = 1$$

which is the equation for an ellipse.

Summary. The identity $\cos^2 t + \sin^2 t = 1$ is an extremely powerful identity, and possibly the one used most often. Here it was used to change from the parametric equation to the equivalent equation using rectangular coordinates.

Example 2: A Polar Graph

Sketch the graph of the polar equation $r = 3\cos\theta$.

Strategy. Again, we start with a table of values. Then we carefully plot the points. Remember that r denotes the distance from the origin, and θ is the angle with the positive x-axis.

Solution.

θ	$r = 3\cos\theta$
0	3
$\frac{\pi}{4}$	$3 \cdot \frac{\sqrt{2}}{2} \approx 2.1$
$\frac{\pi}{2}$	0
$\frac{3\pi}{4}$	$-3 \cdot \frac{\sqrt{2}}{2} \approx 2.1$
π	-3

In order to plot the points, sketch the lines for the different angles. For example, the point $(\frac{\pi}{4}, 2.1)$ is on the ray that makes an angle of $\frac{\pi}{4}$ with the positive x-axis with a distance of 2.1. Note that the ray $\theta = \frac{\pi}{4}$ is the same line as the diagonal $y = x$.

The point $(\pi, -3)$ is on the ray that makes an angle of π with the positive x-axis. Note that this ray is the negative x-axis. The distance is -3, which takes us in the opposite direction of the ray, i.e. it brings us back to the positive x-axis. Thus we get the point $x = +3, y = 0$. Note that this is also our starting point when $\theta = 0$.

Here is the graph of $r = 3\cos\theta$.

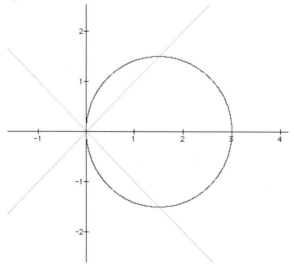

If you use a calculator, then first put the calculator into polar mode. Enter $r = 3\cos\theta$.

Summary. Now that we have technology to help us, graphing in polar coordinates is actually quite fun. The graph here, a circle, is quite interesting coming from such a simple looking equation.

Example 3: Converting Polar to Rectangular Coordinates

Write the equation from example 2 in rectangular coordinates. The equation is $r = 3\cos\theta$.

Strategy. Use the equations
$$x = r\cos\theta$$
$$y = r\sin\theta$$
$$x^2 + y^2 = r^2$$

Solution.
$$r = 3\cos\theta \quad \text{(multiply by } r\text{)}$$
$$r^2 = 3r\cos\theta$$
$$\text{(substitute } x^2 + y^2 = r^2 \text{ and } x = r\cos\theta\text{)}$$
$$x^2 + y^2 = 3x$$

This is actually the equation of the circle, $\left(x - \frac{3}{2}\right)^2 + y^2 = \frac{9}{4}$. See picture in example 2.

Summary. Note once again, how important the Pythagorean Theorem is and how often it is used. This is quite amazing given that Pythagoras lived at least 2000 years ago.

Example 4: Converting Rectangular to Polar Coordinates

Find a polar equation corresponding to $x^2 + y^2 = 4$.

Strategy. Once again we use the equations $x = r\cos\theta$ and $y = r\sin\theta$.

Solution.
$$\begin{aligned} x^2 + y^2 &= 4 \\ r^2\cos^2\theta + r^2\sin^2\theta &= 4 \\ r^2(\cos^2\theta + \sin^2\theta) &= 4 \\ r^2 &= 4 \\ r &= \pm 2 \end{aligned}$$

Summary. Watch the calculator draw the graph. In both cases $r = 2$ or $r = -2$, we end up with a circle centered at $(0,0)$ with radius 2.

For $r = 2$, the graph begins at $(2,0)$ and goes around counterclockwise.
For $r = -2$, the graph begins at $(-2,0)$ and goes around counterclockwise.

CHAPTER 1

Limits and Continuity

1.1. A Brief Preview of Calculus

Key Topics.
- Estimating the Slope of a Curve
- Estimating the Length of a Curve

Worked Examples.
- Estimating the Slope of a Parabola at a Point
- Estimating the Length of a Parabola

Overview

We are now ready to begin our study of calculus. Pretty exciting - it took a lot of years to get here! Calculus deals with quantities that change (like velocity, profit, cost etc.) and we can tell the way a graph changes by looking at the slope of lines tangent to the graph. A line tangent to a curve at a point touches the curve at exactly one point.

Estimating the Slope of a Curve. Once we know two points of a line, we can find the slope of that line using the definition of slope The letter which we use to stand for slope is the letter m.

$$m = \frac{\text{change in } y}{\text{change in } x} = \frac{y_2 - y_1}{x_2 - x_1}$$

This formula gives the slope of the line between the two points (x_1, y_1) and (x_2, y_2). If we have a line through a point (x_1, y_1) and a point (x_2, y_2) close to the point (x_1, y_1) on the graph of a curve and we can find the slope of that line, then we can use that value to estimate the slope of the curve at the point (x_1, y_1). A line through two points on a graph of a curve is called a secant line.

Estimating the Length of a Curve. Think of the curve as being made up of lots of little lines tangent to the curve at consecutive points. It you have a graph of a continuous curve on a graphing calculator, then you can ZOOM IN repeatedly until that portion of the curve appears to be linear (i.e. it looks like a straight line). We can find the length of any line segment using the distance formula:
$$d = \sqrt{(x_2 - x_1)^2 + (y_2 - y_1)^2}$$
gives the distance between the two points (x_1, y_1) and (x_2, y_2).

Example 1: Estimating the Slope of a Parabola at a Point

Estimate the slope of $f(x) = 2x^2 - 5$ at $x = 2$.

Strategy. We will take a point close to $(2, 3)$ and find the slope of the line between that point and $(2, 3)$. We will then take a point closer to $(2, 3)$ and repeat the process. We will do this several times, using points to the right of $(2, 3)$ and then points to the left of $(2, 3)$. A clear pattern will emerge which will enable us to estimate the slope of the line tangent to the graph at $(2, 3)$.

Solution. a) The first point we are going to choose occurs at $x = 3$. $f(3) = 2(3)^2 - 5 = 13$. We need to find the slope of the line between the two points $(2, 3)$ and $(3, 13)$.

$$m = \frac{y_2 - y_1}{x_2 - x_1} = \frac{13 - 3}{3 - 2} = 10$$

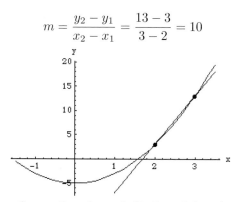

Secant line through $(2, 3)$ and $(3, 13)$

b) The next point we are going to choose occurs at $x = 2.1$. $f(2.1) = 2(2.1)^2 - 5 = 3.82$. We need to find the slope of the line between the two points $(2, 3)$ and $(2.1, 3.82)$.

$$m = \frac{3.82 - 3}{2.1 - 2} = \frac{0.82}{0.1} = 8.2$$

c) Taking one more point to the right of $x = 2$ and even closer to it, we use 2.01. $f(2.01) = 2(2.01)^2 - 5 = 3.0802$, We need to find the slope of the line between the two points $(2, 3)$ and $(2.01, 3.0802)$.

$$m = \frac{3.0802 - 3}{2.01 - 2} = \frac{0.0802}{0.01} = 8.02$$

All of these points were points on the graph to the right of $(2, 3)$. We also need to investigate the slope of secant lines for points to the left of $(2, 3)$.

d) We will start as we did before, with a point one unit away from $x = 2$. Let $x = 1$. Then $f(1) = 2(1)^2 - 5 = -3$. We need the slope of the line between the two points $(2, 3)$ and $(1, -3)$.

$$m = \frac{-3-3}{1-2} = \frac{-6}{-1} = 6$$

e) Next, let $x = 1.9$. Since $f(1.9) = 2.22$,

$$m = \frac{2.22 - 3}{1.9 - 2} = \frac{-0.78}{-0.1} = 7.8$$

f) For our last point let $x = 1.99$. Since $f(1.99) = 2.9202$,

$$m = \frac{2.9202 - 3}{1.99 - 2} = \frac{-0.798}{-0.01} = 7.98$$

Since these slopes approach the number 8, our estimate for the slope of the tangent line is 8.

Summary. As you have undoubtedly noticed, this process is extremely tedious. As we progress, and especially in chapter 3, there will be a much easier way to estimate the slope of a curve. But, what always remains the same, is the very important role the slope plays in the study of calculus.

Example 2: Estimating the Length of a Parabola

Estimate the length of the parabola $f(x) = 2x^2 - 5$ for $1 < x < 3$ by using four line segments.

Strategy. To find the length of a curve on an interval, we will divide the interval into line segments. We can find the length of each line segment and then add them all together.

To understand the concept of the length of a curve, picture forming the curve with a piece of string. Then we can mark the endpoints of the part of the curve which lies on the given interval, straighten out the string and measure the length.

Solution. If we evenly divide the interval $(1, 3)$ into four segments, then the x−values are $x = 1$, $x = 1.5$, $x = 2$, $x = 2.5$, and $x = 3$. The corresponding points on the graph are found by finding $f(1)$, $f(1.5)$, $f(2)$, $f(2.5)$, and $f(3)$ giving us the four points on the curve:

$A(1, -3)$, $B(1.5, -0.5)$, $C(2, 3)$, $D(2.5, 7.5)$, and $E(3, 13)$.

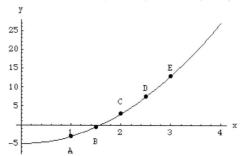

The length of the parabola for $2 < x < 4$ is approximately equal to the sum of the distances between the points A and B, and B and C, and C and D, and D and E. Adding up these distances we get:

$$\begin{aligned}
\text{Length of the curve} &\approx \sqrt{(1.5-1)^2 + (-0.5-(-3))^2} && (\text{distance } \overline{AB}) \\
&+ \sqrt{(2-1.5)^2 + (3-(-0.5))^2} && (\text{distance } \overline{BC}) \\
&+ \sqrt{(2.5-2)^2 + (7.5-3)^2} && (\text{distance } \overline{CD}) \\
&+ \sqrt{(3-2.5)^2 + (13-7.5)^2} && (\text{distance } \overline{DE}) \\
&= \sqrt{16.5} + \sqrt{12.5} + \sqrt{20.5} + \sqrt{30.5} \\
&\approx 16.1354
\end{aligned}$$

Summary. As stated in the summary for example 1, as we go further into the course, we will derive formulas to use to greatly shorten the work needed to solve this type of problem.

1.2. The Concept of Limit

Key Topics.
- Limits
- One Sided Limits
- Determining Limits Graphically and Numerically
- When Limits Fail to Exist

Worked Examples.
- A Piecewise Defined Function
- A Limit Where a Factor Cancels
- A Limit Involving an Absolute Value
- Finding a Limit Graphically and Numerically
- A Couple of Limits That Do Not Exist

Overview

In this section, we are going to talk about limits. Limits underlie almost everything else we study in calculus. It is truly the most basic of the topics.

Limits. You have been using limits all your life, so you really have a basic understanding of what the word means. For example, let's think of a speed limit and what it literally means. If the speed limit says 55 m.p.h., that means that strictly speaking, you can go 54.99 m.p.h., but you cannot go 55.01 m.p.h., right? You can get closer and closer to 55 but you may NOT exceed it. This is one example of a limit.

One Sided Limits. In the speed limit example, we approached 55 from one side only. We can go 54, 54.9, 54.99 etc. until we hit 55 m.p.h. On a graph or number line these numbers are to the left of 55 and we say we are approaching 55 **from the left.** There is a similar definition for approaching a number **from the right.**

Determining Limits Graphically and Numerically. If we look at a graph, we can imagine a little insect crawling along the track made by the graph. The insect gets closer and closer to whatever x-value we are approaching. The y-value of that point is the limit.

If we are looking at a table of values of x which are approaching a certain x-value, we look to see the y-value which the function approaches.

When Limits Fail to Exist. When we approach a given x-value we need to approach it from the left and from the right. If these values are the same, then we call that value **the** limit. However, if the limit from the left and from the right are not the same, then we say that the limit fails to exist.

We can also say that the limit fails to exist if the graph goes off to $\pm\infty$ as we approach the given x-value.

Example 1: A Piecewise Defined Function

For the function $y = f(x)$ graphed below, find the left and right one-sided limits at $x = 0, 1,$ and 3. Decide whether or not the limits exist at these points.

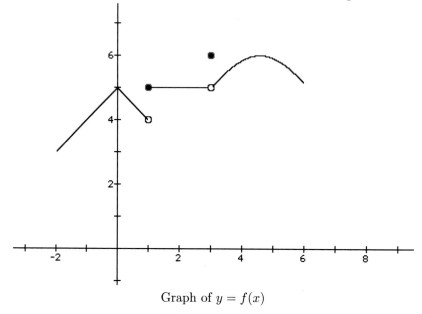

Graph of $y = f(x)$

Strategy. As an insect would, we will *walk* along the graph towards the point with the given x-coordinate. If we get to the same point approaching from the left and from the right, then the conclusion is that the limit exists and has the same y-value. If we get to two different points approaching from the left and the right, then we will conclude that the limit does not exist.

Solution. a) At $x = 0$:
$$\lim_{x \to 0^-} f(x) = 5 \text{ and } \lim_{x \to 0^+} f(x) = 5$$
Since both one-sided limits are equal, the limit at $x = 0$ exists, and $\lim_{x \to 0} f(x) = 5$.

b) At $x = 1$:
$$\lim_{x \to 1^-} f(x) = 4 \text{ and } \lim_{x \to 1^+} f(x) = 5$$
Since both one-sided limits are not equal, the limit at $x = 1$ does not exist, i.e. $\lim_{x \to 1} f(x)$ does not exist.

c) At $x = 3$:
$$\lim_{x \to 3^-} f(x) = 5 \text{ and } \lim_{x \to 3^+} f(x) = 5$$

Since both one-sided limits are equal, the limit at $x = 3$ exists, and $\lim\limits_{x \to 3} f(x) = 5$.

Summary. If there is still some confusion, let your fingers walk along the graph closer and closer to the point with the given x-coordinate and see what y-value you approach. If the answer is the same from both directions, then that is the limit.

You can think of two people building a train track coming towards the middle from either sides. If the tracks *meet* in the middle, then the limit exists. If they do not meet, then the limit does not exist.

Example 2: A Limit Where a Factor Cancels

Evaluate
$$\lim_{x \to 2} \frac{x^2 - 5x + 6}{x^2 - 6x + 8}$$

Strategy. If we substitute 2 in for x, we end up with $\frac{0}{0}$ which is undefined. However, if we use algebra and factor and reduce, then we will "remove" the factor which makes the denominator equal to zero. Then we can substitute 2 for x and find the limit.

Solution.
$$\lim_{x \to 2} \frac{x^2 - 5x + 6}{x^2 - 6x + 8} = \lim_{x \to 2} \frac{(x-2)(x-3)}{(x-2)(x-4)} = \lim_{x \to 2} \frac{(x-3)}{(x-4)} = \frac{-1}{-2} = \frac{1}{2}$$

Summary. When looking for a limit, first try to substitute the number. If the result is $\frac{0}{0}$, then try to factor and reduce and see if the number can then be substituted to find the limit.

Example 3: A Limit Involving an Absolute Value

Evaluate or explain why the limit does not exist.
$$\lim_{x \to 1} \frac{|x-1|}{x-1}$$

Strategy. An absolute value problem always has to be separated into two parts. When the expression inside the absolute value signs is positive, then the expression is unchanged when the absolute value signs are removed. However, when the expression inside the absolute value signs is negative, then the result is the negative of the expression when the absolute value lines are removed. For example: $\mid 2 \mid = 2$ (what was inside the absolute value lines is unchanged), but $\mid -2 \mid = -(-2) = 2$ (the answer is the negative of what was inside the absolute value lines). Consequently, getting rid of the absolute value signs and separating the problem into two parts is the first thing that must be done.

Solution. First note that the absolute value of $x - 1$ can be rewritten as follows:
$$|x-1| = \begin{cases} x-1 & \text{for } x \geq 1 \\ -(x-1) & \text{for } x < 1 \end{cases}$$

From the left:
$$\lim_{x \to 1^-} \frac{|x-1|}{x-1} = \lim_{x \to 1^-} \frac{-(x-1)}{x-1} = -1$$

From the right:
$$\lim_{x \to 1^+} \frac{|x-1|}{x-1} = \lim_{x \to 1^+} \frac{x-1}{x-1} = 1$$

Since both one-sided limits are not equal, the limit at $x = 1$ does not exist. We can write $\lim_{x \to 1} \frac{|x-1|}{x-1}$ does not exist.

Summary. It is sometimes difficult to understand how to separate the function to get rid of the absolute value signs. The hard part is the part where the answer is the negative of what was inside the absolute value signs. In the above example, if $x < 1$, then $x - 1$ is a negative number. Therefore, the opposite, which is $-(x-1)$ is a positive number. For example, if $x = \frac{1}{2}$ then $\frac{1}{2} - 1 = \frac{-1}{2}$ and $\mid \frac{-1}{2} \mid = +\frac{1}{2}$. You just have to think of it in the abstract, which makes it a lot more challenging.

Example 4: Finding a Limit Graphically and Numerically

Use numerical and graphical evidence to determine whether or not the limit exists. If it exists, then state its value.

$$\lim_{x \to 0} \frac{1 - \cos x}{x}$$

Strategy. Sometimes there are no algebraic processes we can use after we get $\frac{0}{0}$ by substituting. We can look at a graph or a table and from that come up with a reasonable conclusion as to what the value of the limit actually is. This is not as accurate or definite as other ways, but sometimes it is the only method we have.

Solution. First numerically:

x	$\frac{1-\cos x}{x}$
1	0.459698
0.1	0.049958
0.01	0.004999
0.001	0.000499

x	$\frac{1-\cos x}{x}$
-1	-0.459698
-0.1	-0.049958
-0.01	-0.004999
-0.001	-0.000499

Since these numbers are getting closer and closer to zero, a reasonable conclusion is that the limit is zero.

Let us check if we can confirm this conjecture graphically:

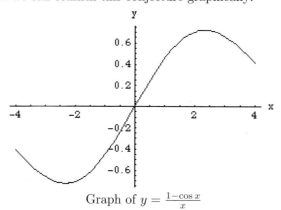

Graph of $y = \frac{1-\cos x}{x}$

When x is approaching zero, so is the graph. Thus from the graph also, we can conclude that

$$\lim_{x \to 0} \frac{1 - \cos x}{x} = 0$$

Summary. Looking at a graph or a table is by no means a certainty, but when we have no other methods at our disposal, then we may have to rely on the "educated" conclusion we can get using them.

Example 5: A Couple of Limits That Do Not Exist

Explain why the following limits do not exist.

a) $\lim\limits_{x \to 1^-} \dfrac{2}{1-x}$

b) $\lim\limits_{x \to 0} \sin\left(\dfrac{1}{x}\right)$

Strategy. Note that when we substituted the values into some of the functions in the earlier examples, we got $\frac{0}{0}$ and proceeded to algebraically manipulate the function so that it was in a form where we could substitute and get an answer. If the numerator is not 0, but the denominator is, then there is usually no way to change the form. In those cases the limit is approaching $\pm\infty$. We will discuss limits involving infinity in a later section.

Sometimes a function, especially a trigonometric function, may jump up and back between numbers. For example, no matter what values we take for $\cos\theta$ or $\sin\theta$, the values keep repeating and we keep getting answers between -1 and $+1$. We will see a picture of this in the second example of this problem. We call the function an **oscillating** function when it keeps jumping up and back and up and back forever.

Solution. a) $f(x) = \dfrac{2}{1-x}$

When plugging in $x = 1$, we get $\frac{2}{0}$ which is undefined. Let us investigate this limit graphically:

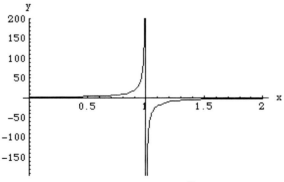

Graph of $f(x) = \dfrac{2}{1-x}$

At $x = 1$, the graph is exploding. When x approaches 1 from the left, the graph is not approaching a number, therefore the limit does not exist. However, we will learn in section 1.4 that this left limit takes off to $+\infty$. Similarly, when x approaches 1 from the right, the graph takes off to $-\infty$, and thus the limit does not exist.

b) $f(x) = \sin\left(\frac{1}{x}\right)$. Let us investigate this limit graphically also.

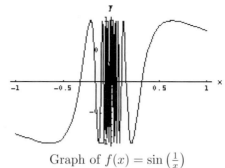

Graph of $f(x) = \sin\left(\frac{1}{x}\right)$

As x approaches zero, $\frac{1}{x}$ is getting larger in magnitude (if x is positive, $\frac{1}{x}$ is very large; if x is negative, $\frac{1}{x}$ is a negative number that is large in magnitude). $\sin\left(\frac{1}{x}\right)$ is oscillating between -1 and 1 as x approaches zero. The closer x gets to zero, the more the graph is oscillating between -1 and 1. Hence the graph does not *settle down* and approach a point, and the limit does not exist.

Summary. We have seen two examples of limits which do not exist - one went off to infinity and the other oscillated. As we progress through the course, there will be many other cases of limits which do not exist.

1.3. Computation of Limits

Key Topics.
- The Limit Laws
- Limits of Known Functions
- Squeeze Theorem

Worked Examples.
- Finding the Limit of Polynomials and Rational Functions
- Finding the Limit of a Rational Function by Canceling a Factor
- Finding a Limit by Rationalizing the Numerator
- A Limit Involving Exponential and Logarithmic Functions
- Finding Limits of a Piece-Wise Defined Function
- A Couple of Squeeze Theorem Examples

Overview

In this section, we will use all the laws of limits with the functions that have been introduced in previous math courses. We will also introduce a theorem with a very strange name: The Squeeze Theorem also known as The Sandwich Theorem. The names make perfect sense in describing what they say, but they definitely have the weirdest names of all the theorems.

The Limit Laws. The laws which govern limits basically tell us that we can do exactly what we think we should be able to do with limits. They say, basically, that the limit of the (sum, difference, quotient, product or constant multiple) is the (sum, difference, quotient, product or constant multiple) of the limit. For example,

$$\text{if } \lim_{x \to a} f(x) = L_1 \text{ and } \lim_{x \to a} g(x) = L_2$$
$$\text{then } \lim_{x \to a} [f(x) + g(x)] = L_1 + L_2$$

Limits of Known Functions. At this point, there are many functions with which we have worked a lot in previous courses and thus "know". These include polynomial, rational, nth root, exponential, logarithmic, trigonometric, and inverse trigonometric functions. We will work with some of these types of functions in this section.

Squeeze Law. If there is a function and we know the smallest and largest values that it can ever take, then we can "squeeze" it between those two values. A great example is $\sin \theta$ which is never less than -1 nor greater than $+1$. If we can squeeze a function between two functions both of which are approaching the same limit, then we can conclude that the given function is approaching that limit also, since it is **always** between the other two.

Example 1: Finding the Limit of Polynomials and Rational Functions

---

Evaluate a) $\lim_{x \to 2} (2x^2 - 3x + 2)$, and b) $\lim_{x \to 1} \frac{2x^2 - 5}{2x + 1}$.

Strategy. As in the previous section, if we can substitute the number into the function to find the limit, then we will do that.

Solution. a)
$$\lim_{x \to 2} (2x^2 - 3x + 2) = 2 \cdot 2^2 - 3 \cdot 2 + 2 = 4$$

b)
$$\lim_{x \to 1} \frac{2x^2 - 5}{2x + 1} = \frac{2 \cdot 1^2 - 5}{2 \cdot 1 + 1} = \frac{-3}{3} = -1$$

Summary. As we move ahead into more complicated functions, remember that the first process to try is to simply substitute. In other words, see if you can get the answer by using
$$\lim_{x \to a} f(x) = f(a)$$
If that is not possible, then try something else.

Example 2: Finding the Limit of a Rational Function by Canceling a Factor

Evaluate
$$\lim_{x \to 1} \frac{3x^2 - 3x}{x^2 - 1}$$

Strategy. If we try to substitute 1 for x in $\frac{3x^2-3x}{x^2-1}$, we get $\frac{0}{0}$ which is undefined. With a rational algebraic expression, try factoring as the next step.

Solution.
$$\lim_{x \to 1} \frac{3x^2 - 3x}{x^2 - 1} = \lim_{x \to 1} \frac{3x(x-1)}{(x+1)(x-1)} = \lim_{x \to 1} \frac{3x}{x+1} = \frac{3}{2}$$

Summary. Whenever, the result of substitution is $\frac{0}{0}$, then there is something which can be done. Often, factoring is what works.

Example 3: Finding a Limit by Rationalizing the Numerator

Evaluate
$$\lim_{x \to 1} \frac{\sqrt{x+3} - 2}{x - 1}$$

Strategy. In algebra we often rationalized a denominator in order to avoid having radical signs in the denominator of a fraction. In calculus we often rationalize the numerator in an attempt to find a limit which resulted in $\frac{0}{0}$ when we tried to substitute.

Solution.
$$\begin{aligned}
\lim_{x \to 1} \frac{\sqrt{x+3} - 2}{x - 1} &= \lim_{x \to 1} \frac{\left(\sqrt{x+3} - 2\right)}{(x-1)} \cdot \frac{\left(\sqrt{x+3} + 2\right)}{\left(\sqrt{x+3} + 2\right)} \\
&= \lim_{x \to 1} \frac{(x+3) + 2\sqrt{x+3} - 2\sqrt{x+3} - 4}{(x-1)\left(\sqrt{x+3} + 2\right)} \\
&= \lim_{x \to 1} \frac{x + 3 - 4}{(x-1)\left(\sqrt{x+3} + 2\right)} \\
&= \lim_{x \to 1} \frac{x - 1}{(x-1)\left(\sqrt{x+3} + 2\right)} \\
&= \lim_{x \to 1} \frac{1}{\left(\sqrt{x+3} + 2\right)} \\
&= \frac{1}{\left(\sqrt{1+3} + 2\right)} = \frac{1}{4}
\end{aligned}$$

Summary. Although it seems strange at first because it is a new idea, rationalizing the numerator is frequently used as a means of simplifying an expression.

Example 4: A Limit Involving Exponential and Logarithmic Functions

Evaluate $\lim_{x \to 1} \left(\ln(2-x) + 3e^{1-x} - 2 \right)$.

Strategy. Once again, the first method to try is substituting. In this problem, it works!

Solution.
$$\begin{aligned}
\lim_{x \to 1} &\left(\ln(2-x) + 3e^{1-x} - 2 \right) \\
&= \ln(2-1) + 3e^{1-1} - 2 \\
&= \ln 1 + 3e^0 - 2 \\
&= 0 + 3 - 2 \\
&= 1
\end{aligned}$$

Summary. As stated before, if substitution can be done, that is what we do.

Example 5: Finding Limits of a Piece-Wise Defined Function

Evaluate the following limits for the function $f(x)$ defined below.
a) $\lim_{x \to 0^-} f(x)$, $\lim_{x \to 0^+} f(x)$, and $\lim_{x \to 0} f(x)$,
b) $\lim_{x \to 2^-} f(x)$, $\lim_{x \to 2^+} f(x)$, and $\lim_{x \to 2} f(x)$.

$$f(x) = \begin{cases} 2x^2 + 1 & \text{for } x \leq 0 \\ \cos(\pi x) & \text{for } 0 < x < 2 \\ 2x - 2 & \text{for } x \geq 2 \end{cases}$$

Strategy. It helps here to actually sketch a number line, because we need to see the relationship between numbers.

We will check the limits from the right and from the left. If they are the same, then the limit exists and has the same value as the two one-sided limits.

Solution. a) For $\lim_{x \to 0^-} f(x)$ we need $x < 0$, so we use $f(x) = 2x + 1$.

For $\lim_{x \to 0^+} f(x)$ we need $x > 0$, so we need the piece where $0 < x < 2$, and so we use $f(x) = \cos(\pi x)$.

Continue in that manner.

$$\lim_{x \to 0^-} f(x) = \lim_{x \to 0^-} (2x^2 + 1) = 1$$

$$\lim_{x \to 0^+} f(x) = \lim_{x \to 0^+} \cos(\pi x) = \cos 0 = 1$$

$$\lim_{x \to 0} f(x) = 1$$

b)

$$\lim_{x \to 2^-} f(x) = \lim_{x \to 2^-} \cos(\pi x) = \cos(2\pi) = 1$$

$$\lim_{x \to 2^+} f(x) = \lim_{x \to 2^+} 2x - 2 = 2$$

$$\lim_{x \to 2} f(x) \text{ does not exist}$$

The limit at $x = 2$ does not exist, because the two one-sided limits do not match.

We can double check our numbers by looking at the graph of $y = f(x)$.

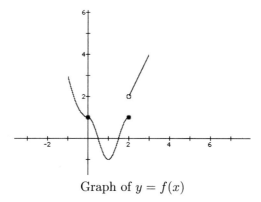

Graph of $y = f(x)$

Summary. Looking at the graph really helps us see what the graph is approaching as we come from the left and from the right. If we approach the same point coming from either direction, whether we actually touch it or not, that is the limit.

Done analytically, decide which piece of the function is needed and then substitute the number into that part of the function.

Example 6: A Couple of Squeeze Theorem Examples

Determine the limits: a) $\lim\limits_{x \to 0} x^2 \sin\left(\frac{1}{x^2}\right)$, and b) $\lim\limits_{x \to 0} x \cos\left(\frac{1}{x}\right)$.

Strategy. The Squeeze Theorem (which is sometimes called The Sandwich Theorem) is used often when working with the sine and cosine functions. The reason is that both of these functions are squeezed or sandwiched between -1 and $+1$.

Solution. First we investigate these limits graphically:

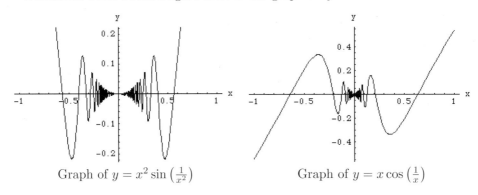

Graph of $y = x^2 \sin\left(\frac{1}{x^2}\right)$ Graph of $y = x \cos\left(\frac{1}{x}\right)$

Both of these graphs are oscillating, but the oscillations are getting smaller and smaller as $x \to 0$. This leads us to expect that the limit exists. Let's use The Squeeze Theorem and see if we can show this analytically.

The following inequalities hold true for all angles θ.
$$-1 \leq \sin \theta \leq 1 \quad \text{and} \quad -1 \leq \cos \theta \leq 1$$
We can use this inequality to *squeeze* our functions:

a) In the first example, $\theta = \frac{1}{x^2}$
$$-1 \leq \sin \theta \leq 1 \implies -1 \leq \sin\left(\frac{1}{x^2}\right) \leq 1$$

We can multiply this inequality by x^2, since $x^2 > 0$ for all x.
$$-x^2 \leq x^2 \sin\left(\frac{1}{x^2}\right) \leq x^2$$

Applying the squeeze theorem, we get:
$$\lim_{x \to 0} \left(-x^2\right) \leq \lim_{x \to 0} \left[x^2 \sin\left(\frac{1}{x^2}\right)\right] \leq \lim_{x \to 0} \left(x^2\right)$$
$$0 \leq \lim_{x \to 0} \left[x^2 \sin\left(\frac{1}{x^2}\right)\right] \leq 0$$

The limit we are looking for is "squeezed" between two functions which are both approaching zero. Therefore, our function is going to zero also. (It is like the turkey in between two slices of bread. The turkey goes wherever the bread goes.)

Therefore, we can state that:
$$\lim_{x \to 0} x^2 \sin\left(\frac{1}{x^2}\right) = 0$$

b)
$$-1 \leq \cos\left(\frac{1}{x}\right) \leq 1$$

In this example we cannot simply multiply the inequality by x, since x can be positive or negative. We will, therefore, work it in two steps - one with $\dot{x} > 0$ and one with $x < 0$.

Let's start with what we know:
$$-1 \leq \cos\theta \leq 1 \implies -1 \leq \cos\left(\frac{1}{x}\right) \leq 1$$

If $\dot{x} > 0$:
$$-x \leq x\cos\left(\frac{1}{x}\right) \leq x$$

and also note that
$$\lim_{x \to 0}(-x) = 0 \quad \text{and} \quad \lim_{x \to 0}(x) = 0$$

Since $x\cos\left(\frac{1}{x}\right)$ is *squeezed* between two functions both approaching zero, we can conclude that:
$$\lim_{x \to 0} x\cos\left(\frac{1}{x}\right) = 0$$

If $x < 0$:
$$-x \geq x\cos\left(\frac{1}{x}\right) \geq x$$
$$x \leq x\cos\left(\frac{1}{x}\right) \leq -x$$

Then using the same reasoning as we did for $x > 0$, we can conclude again that:
$$\lim_{x \to 0} x\cos\left(\frac{1}{x}\right) = 0$$

Summary. When using inequalities it is important to remember that if you multiply or divide by a negative number, the direction of the inequality reverses.

1.4. Continuity And Its Consequences

Key Topics.
- Continuous Functions
- Intermediate Value Theorem

Worked Examples.
- Determining Continuity from a Graph
- Determining Continuity of a Piece-Wise Defined Function
- A Rational Function with a Removable Discontinuity
- Showing the Existence of Zeros of a Polynomial

Overview

This section deals with continuity. Basically, this is a very intuitive topic. It means exactly what it means in normal conversation. The graph can be drawn without any breaks of any kind. It can be drawn without taking the pencil off of the paper.

Continuous Functions. Continuity must be formally defined so that it is always possible to determine whether or not a function is continuous at any given point. In order for a function to be continuous at a point, it must be defined at that point and it must have a limit at that point. In addition, the value of the function at the point must be **equal** to the limit at the point.

Intermediate Value Theorem. This is another theorem whose result does not surprise us, although the actual proof is usually not seen until a more advanced calculus course. It says that if a function is continuous on an interval $[a,b]$ then the function takes on every value between $f(a)$ and $f(b)$. Think about what this says. If, for example, the function is continuous on $[a,b]$ and $f(a) = 3$ and $f(b) = 5$, then the function will take on every value between 3 and 5. This should make sense because since it is continuous it has to get from $y = 3$ to $y = 5$ without jumping or skipping over any point, it **must** hit every value between 3 and 5.

Example 1: Determining Continuity from a Graph

The graph of $y = f(x)$ is shown below. Is f continuous at $x = 0$, $x = 1$, and $x = 3$?

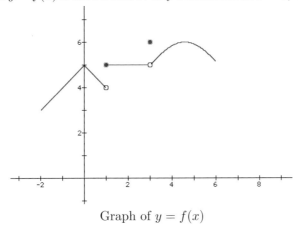

Graph of $y = f(x)$

Strategy. We can analyze the graph at each of the points to determine whether the function value is equal to the limit at that point. If it is, then we conclude that the function is continuous at the given point.

Solution. a) f is continuous at $x = 0$, since $\lim\limits_{x \to 0} f(x) = f(0)$

b) f is not continuous at $x = 1$, or $x = 3$.

At $x = 1$, the graph of f has a jump and hence the limit does not exist at that point.

At $x = 3$, f has a removable discontinuity. The limit exists at the point, and the function exists at the point, but they have two different values. $\lim\limits_{x \to 3} f(x) = 5$ but $f(3) = 6$.

Summary. Hopefully, the definition of continuity makes more sense here when you have a graph to look at.

Example 2: Determining Continuity of a Piece-Wise Defined Function

Determine the intervals where $f(x) = \begin{cases} 2x^2 + 1 & \text{for } x \leq 0 \\ \cos(\pi x) & \text{for } 0 < x < 2 \\ 2x - 2 & \text{for } x \geq 2 \end{cases}$ is continuous.

Strategy. We will look analytically at this function. Once again, we need to see if the function value is equal to the limit at a given point. The graph which is sketched will support our conclusions.

Solution. We only need to see whether f is continuous at $x = 0$, and $x = 2$, since f is continuous for all $x \neq 0$ and $x \neq 2$,

$$\lim_{x \to 0^-} f(x) = \lim_{x \to 0^-} (2x^2 + 1) = 1$$
$$\lim_{x \to 0^+} f(x) = \lim_{x \to 0^+} \cos(\pi x) = \cos 0 = 1$$
$$\lim_{x \to 0} f(x) = 1 = f(0)$$

Therefore f is continuous at $x = 0$.

$$\lim_{x \to 2^-} f(x) = \lim_{x \to 2^-} \cos(\pi x) = \cos(2\pi) = 1$$
$$\lim_{x \to 2^+} f(x) = \lim_{x \to 2^+} (2x - 2) = 2$$
$$\lim_{x \to 2} f(x) \text{ does not exist}$$

Therefore f is not continuous at $x = 2$.

We conclude that f is continuous for all real numbers except for $x = 2$, that is f is continuous on the intervals $(-\infty, 2) \cup (2, \infty)$. We can also see this from the graph:

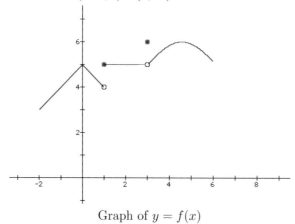

Graph of $y = f(x)$

Summary. In the previous two examples, we determined continuity analytically and graphically. The answers to any problem must match no matter which method is used to solve it.

Example 3: A Rational Function with a Removable Discontinuity

Find the intervals where f is continuous.
$$f(x) = \frac{x^2 - 4}{2x - 4}$$

Strategy. The difficulty in this problem arises from the fact that the function is defined everywhere except at $x = 2$. If the function does not exist at a point, in this case at $x = 2$, then it cannot be continuous at that point.

Solution. For the solution, we will use the discontinuous function as it was originally given.

$f(x) = \frac{x^2-4}{2x-4}$ is defined for all $x \neq 2$, and continuous for all $x \neq 2$. Is f continuous at $x = 2$?

$$\lim_{x \to 2} \frac{x^2 - 4}{2x - 4} = \lim_{x \to 2} \frac{(x+2)(x-2)}{2(x-2)} = \lim_{x \to 2} \frac{x+2}{2} = 2$$

Since the limit exists at $x = 2$, but f is undefined at $x = 2$, f is not continuous at $x = 2$. This discontinuity is a removable one, as explained in the strategy section. See the little hole in the graph:

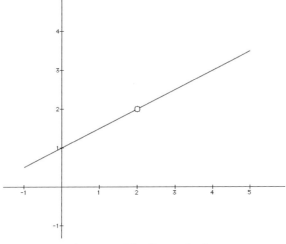

A removable discontinuity

Summary. However, the limit exists and can be found by factoring and cancelling or reducing. The **only** point where this function is discontinuous is at $x = 2$. Consequently, there is a teeny hole in the otherwise continuous graph. This type of discontinuity is called a **removable discontinuity** because with a redefinition of the function it can be made to be continuous. In other words, we can remove the discontinuity by defining $f(x)$ as:
$$f(x) = \begin{cases} \frac{x^2-4}{2x-4} & \text{for } x \neq 2 \\ 2 & \text{for } x = 2 \end{cases}$$
A removable discontinuity looks a little bit like a cavity in a tooth. It is a little hole which can be filled in with the right redefinition of $f(x)$.

1.4. CONTINUITY AND ITS CONSEQUENCES

Example 4: Showing the Existence of Zeros of a Polynomial

Use the Intermediate Value Theorem to show that the polynomial $f(x) = x^3 - 4x + 2$ has three zeros.

Strategy. By The Fundamental Theorem of Algebra we know that there are at most three zeros in a polynomial of degree 3.

Hence, we only need to show that this function does have all three zeros. If we can find an a and a b such that $f(a) < 0$ and $f(b) > 0$ or vice versa, then by the Intermediate Value Theorem, f takes on every value between $f(a)$ and $f(b)$ implying that there exists a c between a and b such that $f(c) = 0$. Our job now is to find the a and b where this occurs.

Solution. We start by evaluating f at "random" points looking for a sign change in consecutive $f(x)$ values.

x	$f(x) = x^3 - 4x + 2$
-3	-13
-2	2
-1	5
0	2
1	-1
2	2

f is a polynomial and therefore a continuous function. By looking at the table of values we can see that f goes from negative to positive between $x = -3$ and $x = -2$, and from positive to negative between $x = 0$ and $x = 1$, and then again from negative to positive between $x = 1$ and $x = 2$.

Since $f(-3) < 0$, and $f(-2) > 0$, by the Intermediate Value, there must be a number c_1 between -3 and -2 with $f(c_1) = 0$. Thus we have one zero between -3 and -2.

Similarly since $f(0) > 0$, and $f(1) < 0$, by the Intermediate Value, there must be a number c_2 between 0 and 1 with $f(c_2) = 0$. Here we have a zero between 0 and 1.

And finally since $f(1) < 0$, and $f(2) > 0$, by the Intermediate Value, there must be a number c_3 between 1 and 2 with $f(c_3) = 0$. The last zero lies between 1 and 2.

Therefore f does have three zeros.

Summary. If you have a graphing calculator, it is easy to find the x-values to use so that the sign of $f(x)$ changes for consecutive x-values. You can see where the graph crosses the x-axis. These x-intercepts are the zeros we are asked to find.

1.5. Limits Involving Infinity

Key Topics.
- Limits That Are Infinite
- Limits at Infinity
- Vertical Asymptotes
- Horizontal Asymptotes
- Slant Asymptotes

Worked Examples.
- A Limit that is Infinite
- Limits at Infinity of Rational Functions
- Limits Involving Exponential and Logarithmic Functions
- Finding Asymptotes of a Rational Function
- A Slant Asymptote

Overview

The definition of limit that we have used requires that the limit be a finite real number. We have seen $\lim_{x \to a} f(x) = L$ where L is a finite real number. However, there are many instances where a limit doesn't exist. If the reason it doesn't exist is because it goes off to infinity, then there is a special notation which gives us a lot more information than just saying that it does not exist.

Limits That Are Infinite. If you try to substitute to obtain the limit and the denominator equals zero, but the numerator is not zero, then we can say that the function "explodes" at this point and goes off to $\pm\infty$. We need to analyze the function at the point to decide whether it approaches $+\infty$ or $-\infty$.

Limits at Infinity. In this section we work with limits where $x \to \infty$ or $x \to -\infty$. We are looking at what happens to the function as we move far away from the $y-$axis. When working with a rational function, a great method to use is to divide all the terms by x to this highest power which appears in the denominator. In this way we are guaranteed a non-zero denominator when we take the limit.

Vertical Asymptotes. You probably learned that in rational functions, there is a vertical asymptote whenever the denominator is equal to zero after the function has been fully reduced. The graph approaches but **never** crosses a vertical asymptote. The function is undefined at that point. On either side of the vertical asymptote the graph approaches $\pm\infty$.

Other examples of functions which have vertical asymptotes, where the function approaches $\pm\infty$ on either side are $f(x) = \ln x$ at $x = 0$, and $f(x) = \tan x$ at $x = \pm\dfrac{\pi}{4}$.

Horizontal Asymptotes. Horizontal asymptotes, $y = c$, are obtained when we find $\lim_{x \to \pm\infty} f(x) = c$. This asymptote may be crossed, but eventually the graph *settles down* and approaches $y = c$ as $x \to \pm\infty$.

We find the horizontal asymptotes by finding $\lim_{x \to \pm\infty} f(x)$. As stated above, if we have a rational function, we divide top and bottom by x to the highest power which appears in the denominator.

Slant Asymptotes. A slant asymptote is a line which the graph approaches which is neither horizontal nor vertical. In a rational function this occurs when the degree of the numerator is exactly one number higher than the degree of the denominator. It is found using long division.

Example 1: A Limit that is Infinite

Evaluate
$$\text{a)} \quad \lim_{x \to 3^-} \frac{2x-2}{x-3} \quad \text{and b)} \quad \lim_{x \to 3^+} \frac{2x-2}{x-3}$$

Strategy. Clearly, if we substitute 3 into this function we end up with 0 in the denominator, but not in the numerator. Therefore, we know that the graph is "exploding" at $x = 3$. Our job is to determine if it is going to $+\infty$ or $-\infty$ on either side of $x = 3$.

Solution. a) Let's look first at
$$\lim_{x \to 3^-} \frac{2x-2}{x-3}$$
Substituting 3 for x results in $\frac{4}{0}$. The numerator, then is positive no matter what value is put in for x. As we approach 3 from the left we have x-values close to but less than 3 (like 2.99). Therefore, when we subtract 3 in the denominator's $x-3$, we have a negative number and we conclude that:
$$\lim_{x \to 3^-} \frac{2x-2}{x-3} = -\infty$$

b) A similar process for
$$\lim_{x \to 3^+} \frac{2x-2}{x-3}$$
has us looking at x-values close to but larger than 3 (like 3.01). Therefore, when we subtract 3 in $x-3$, we have a positive number in the denominator along with a positive number in the numerator, and thus we conclude
$$\lim_{x \to 3^+} \frac{2x-2}{x-3} = +\infty$$

Summary. If there is a vertical asymptote, we need to analyze the sign of the numerator and the sign of the denominator as x approaches the number from the left and from the right. In that way we can determine whether the graph approaches $+\infty$ or $-\infty$.

Example 2: Limits at Infinity of Rational Functions

Evaluate
$$\text{a) } \lim_{x \to \infty} \frac{6x^2 - 2}{2x^2 + 3x} \quad \text{and b) } \lim_{x \to -\infty} \frac{6x^3 - 2}{2x^2 + 3x}$$

Strategy. Notice that the main difference in these two problems is the degree of the numerator. (We go through the exact same process for $x \to +\infty$ as for $x \to -\infty$).

We are going to divide top and bottom by x to the highest power which appears in the denominator. In these problems, we divide by x^2.

Note that $\lim_{x \to \infty} \frac{c}{x^n} = 0$ for c, any constant and n, a positive rational number. What this says is that if the denominator of the fraction is getting huge, and the top is not, then the whole fraction is approaching zero. Think about a number like $\frac{4}{7,000,000}$. Pretty close to zero, don't you agree? (Similar reasoning occurs for $x \to -\infty$.)

Solution. For both limits divide the numerator and denominator by x^2, the largest power of x that appears in the denominator in both examples.

a) Dividing by x^2, we get:
$$\lim_{x \to \infty} \frac{6x^2 - 2}{2x^2 + 3x} = \lim_{x \to \infty} \frac{\frac{6x^2}{x^2} - \frac{2}{x^2}}{\frac{2x^2}{x^2} + \frac{3x}{x^2}} = \lim_{x \to \infty} \frac{6 - \frac{2}{x^2}}{2 + \frac{3}{x}} = \frac{6 - 0}{2 + 0} = 3$$

b) Using a similar method:
$$\lim_{x \to -\infty} \frac{6x^3 - 2}{2x^2 + 3x} = \lim_{x \to -\infty} \frac{6x - \frac{2}{x^2}}{2 + \frac{3}{x}} = \lim_{x \to -\infty} \frac{6x - 0}{2 + 0} = \lim_{x \to -\infty} (3x) = -\infty$$

Summary. The one little difference in the two problems led to quite different answers.

Example 3: Limits Involving Exponential and Logarithmic Functions

Evaluate
$$\text{a) } \lim_{x \to \infty} e^{-x^2} \text{ and b) } \lim_{x \to 1^+} [\ln(x-1)]$$

Strategy. If you know what the graphs of these functions look like, it will help you **see** what is happening. We will use analytic methods here.

Solution. a) If $x \to \infty$, then $x^2 \to \infty$ also, it just goes faster. We have e^{-x^2}, which can be rewritten as $\dfrac{1}{e^{x^2}}$. Since e is a positive number, we have a denominator getting very large with a numerator remaining constant. Hence, we conclude:

$$\lim_{x \to \infty} e^{-x^2} = \lim_{x \to \infty} \frac{1}{e^{x^2}} = 0$$

b) For $\lim_{x \to 1^+}[\ln(x-1)]$, look at some values near 1 and greater than 1.

x	$f(x)$
1.1	-2.3
1.01	-4.6
1.001	-6.9
1.00001	-11.5

If we continue in this manner, we will see that $f(x)$ approaches $-\infty$ as $x \to 1^+$ or

$$\lim_{x \to 1^+} \ln(x-1) = -\infty$$

Summary. If the function is undefined when substitution is tried and we are looking for a limit as $x \to \pm\infty$, we need to do a careful analysis of the function to determine what is happening.

Example 4: Finding Asymptotes of a Rational Function

Find all vertical and horizontal asymptotes of the function

$$f(x) = \frac{2x^2 + 10x}{x^2 - 25}$$

Strategy. To find the vertical asymptote we will see where the denominator is equal to zero after the function has been fully reduced.

To find any horizontal asymptotes we will divide the top and the bottom by x^2 and then find $\lim_{x \to \pm\infty} f(x)$.

Solution. Vertical Asymptotes

$$f(x) = \frac{2x^2 + 10x}{x^2 - 25} = \frac{2x(x+5)}{(x-5)(x+5)} = \frac{2x}{x-5}$$

At $x = -5$, there is a removable discontinuity which looks like a hole in the graph. There is no vertical asymptote at $x = -5$.

At $x = 5$, there is a vertical asymptote. We need to determine what the graph is doing on either side of this asymptote.

Let's investigate the left and right-limits for $x = 5$.

$$\lim_{x \to 5^-} f(x) = \lim_{x \to 5^-} \frac{2x}{x-5}$$

This is similar to example 1 of this section. We are looking at numbers close to 5 but less than 5. The numerator will be positive. We need to determine the sign of the denominator. When $x < 5$, then $x - 5$ is negative and we have:

$$\lim_{x \to 5^-} f(x) = -\infty$$

Similarly when $x > 5$, $x - 5$ is positive, thus

$$\lim_{x \to 5^+} f(x) = +\infty$$

Therefore f has a vertical asymptote at $x = 5$.

Horizontal Asymptotes

In order to find horizontal asymptotes we need to investigate the limits at infinity.

$$\lim_{x \to \pm\infty} \frac{2x}{x-5} = \lim_{x \to \pm\infty} \frac{2}{1 - \frac{5}{x}} = \frac{2}{1-0} = 2$$

Thus the graph of f has a horizontal asymptote at $y = 2$.

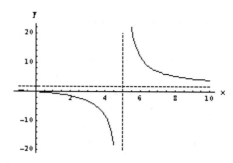

Summary. The graph we have here completely supports the analytical work we did. If there were any discrepancy we would go over our analytical work and/or check how we entered the function into the program to obtain the graph.

Example 5: A Slant Asymptote

Find the slant asymptote of the rational function $f(x) = \dfrac{x^2 + 1}{x - 1}$.

Strategy. We know that there is a slant asymptote here because the numerator is exactly one degree higher than the denominator. We can find the slant asymptote using long division.

Solution. Divide $x^2 + 1$ by $x - 1$.

$$
\begin{array}{r}
x + 1 \\
x - 1 \overline{\smash{\big)} x^2 + 0x + 1} \\
\underline{(x^2 - 1x)} \\
x + 1 \\
\underline{x - 1} \\
\text{Remainder} = 2
\end{array}
$$

Thus
$$f(x) = \frac{x^2 + 1}{x - 1} = x + 1 + \frac{2}{x^2 + 1}$$

When x is approaching $\pm\infty$, $f(x)$ approaches the line $y = x + 1$. Thus f does have a slant asymptote, and the asymptote is given by

$$y = x + 1$$

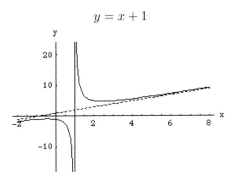

Summary. Once again, the graph supports the analytical work.

1.6. Limits and Loss-Of-Significance Errors

Key Topics.
- Loss of Significance Error

Worked Examples.
- A small Error with a Disastrous Effect
- Investigation of a Loss of Significance Error

Overview

When using calculators or computers, we must remember that we have to also use our brains. When we enter something into a calculator we should have some idea about how the answer should look so that if we accidentally hit a wrong button or forget a parenthesis which changes the calculations, we will realize that there is an error.

In addition to us doing something wrong, there can be a problem caused by the fact that calculators perform computations only approximately. It is true that they perform the calculations to a very high degree of accuracy, but if there are several computations in a row with rounding, the results can become quite disastrous. Look at the graphs and the tables in your text in section 1.6 to see what happens when a function behaves erratically.

Loss of Significance Error. In mathematics we often work with very large or very small numbers. Calculators (and people) round the numbers which may result in a totally incorrect answer. An example follows which shows what may happen when subtracting numbers which have been rounded.

Problems with rounding up numbers and subtracting numbers, such as:
$1.23456 - 1.23457$ if rounded to five digits gives $1.2345 - 1.2345 = 0$
This can cause disastrous events when combined with taking limits.

$$\lim_{x \to 0^+} \frac{(1.23456 - 1.23457)}{x} = -\infty \quad \text{and} \quad \lim_{x \to 0^+} \frac{(1.2345 - 1.2345)}{x} = 0$$

Example 1: A small Error with a Disastrous Effect

Compare the limits to show that small error can have disastrous effects.

$$\text{a) } \lim_{x \to 1^+} \frac{x-1}{x^2-1} \quad \text{and b) } \lim_{x \to 1^+} \frac{x - 0.99999}{x^2 - 1}$$

Strategy. Note that these problems are very similar, yet have completely different results. If we do not round 0.99999, then when we substitute 1 for x, the numerator does not become 0. We have a numerator of -0.000001 over a denominator which is approaching 0, and is positive. Follow the solution and see what happens to the two limits.

Solution. a) In the first limit, we can factor the denominator and then we can cancel the factor $x - 1$.

$$\lim_{x \to 1^+} \frac{x-1}{x^2-1} = \lim_{x \to 1^+} \frac{x-1}{(x+1)(x-1)} = \lim_{x \to 1^+} \frac{1}{x+1} = \frac{1}{2}$$

b) In the second limit, there is nothing to cancel. It is a limit of type $\frac{-0.00001}{0^+}$

$$\lim_{x \to 1^+} \frac{x - 0.99999}{x^2 - 1} = -\infty$$

We see that a small rounding error can significantly change the limit.

Summary. We tend to round frequently. These examples should act as a warning that we should be extremely careful and make sure that we are not changing the problem.

Example 2: Investigation of a Loss of Significance Error

Investigate the loss of significance error in finding the limit

$$\lim_{x \to \infty} \frac{(x+1)^2 - x^2}{x}$$

Strategy. If x is a very large number, then $(x+1)^2$ and x^2 are very close to each other. Subtracting them from one another may give 0 as an answer when rounding has occurred. We will show in this problem that if $x = 10^4 = 1 \cdot 10^4$ the calculation on a calculator is okay. However, when $x = 10^6 = 1 \cdot 10^6$, the calculation on the calculator completely disintegrates because the rounding that the calculator performs causes a major change in the values.

Solution. We first look at the graph. When x is large, here x is in the order of $10^7 = 10,000,000$, the graph starts to wildly oscillate, and we can see a loss of significance error.

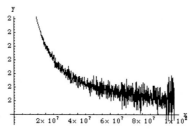

How does this error appear? We are subtracting two numbers, $(x+1)^2$ and x^2 in the numerator. When x is large these two numbers are very close. A calculator with a four digit mantissa would compute this in the following way:

x	$(x+1)^2 - x^2$	Calculator Approximation
10^4	$20,001$	$1.0001^2 \cdot (10^4)^2 - (1 \cdot 10^4)^2 \approx 1.0002 \cdot 10^8 - 10^8 = 20,000$
10^6	$2,000,001$	$1.000001^2 \cdot (10^6)^2 - (1 \cdot 10^6)^2 \approx 10^{12} - 10^{12} = 0$

The approximation for $x = 10^4$ is a very close approximation, but the approximation for $x = 10^6$ is a disastrous approximation.

In order to avoid these rounding errors, we have to avoid subtracting two numbers that are very close. Here we can simplify the function as follows:

$$\lim_{x \to \infty} \frac{(x+1)^2 - x^2}{x} = \lim_{x \to \infty} \frac{x^2 + 2x + 1 - x^2}{x} = \lim_{x \to \infty} \frac{2x+1}{x} = 2$$

Summary. As stated in the text, if at all possible, avoid subtraction of **nearly equal values**. Try to accomplish this avoidance by using algebraic manipulation as shown in this example.

CHAPTER 2

Differentiation

2.1. Tangent Lines and Velocity

Key Topics.
- Secant and Tangent Lines
- Average Rate of Change
- Instantaneous Rate of Change

Worked Examples.
- Finding Average Rates of Change from a Table
- Estimating the Slope of a Tangent Line from a Graph
- Finding Instantaneous Rates of Change of a Polynomial
- Estimating Instantaneous Rates of Change from a Table
- Finding the Equation of a Tangent Line at a Point

Overview

What you are able to do with calculus which you cannot do with only algebra, is deal with change. There are two main branches of calculus - differentiation and integration. The first calculus course concentrates mostly on differentiation which measures the rate of change.

Secant and Tangent Lines. Picture any graph which is not a straight line. For example graph a simple parabola, $y = x^2$.

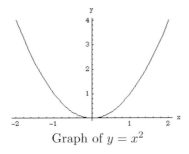
Graph of $y = x^2$

Notice that the graph is decreasing on the left side of the y-axis, increasing on the right side of the y-axis, and not changing at all at $x = 0$. If you want to know how the graph changes (i.e. increasing, decreasing or staying the same) you need to be specific about which part of the graph you are working with.

If you draw a line which touches the graph at *exactly* one point, you can find the slope of that line. You have been finding the slopes of lines since you first studied algebra. That slope will closely approximate the change of the graph at that point. That very important line is called the *tangent* line. Note: a tangent line to a graph touches the graph at exactly ONE point.

The problem is that you need *two* points to find the slope of a line. Remember: The letter used for slope is m.

$$m = \frac{\text{change in } y}{\text{change in } x}$$

or if the two points are (x_1, y_1) and (x_2, y_2), then

$$m = \frac{y_2 - y_1}{x_2 - x_1}.$$

A line which touches the graph in exactly two points is called a *secant* line. We will use limits which you saw in chapter one to go from the slope
of the secant line to the slope of the tangent line.

Average Rate of Change. The average rate of change is actually what you have dealt with most of your life. The key work here is *rate*. If you are driving 50 m.p.h., then 50 m.p.h. is the rate of change. You are moving 50 miles for every hour you drive. So if you drove for 3 hours and covered 150 miles, then you know that on average you went 50 m.p.h. How? The distance in miles per hour is $\frac{\text{number of miles}}{\text{time in hours}}$. This is the familiar $D = RT$ formula when it has been solved for R.

$$R = \frac{D}{T}$$

This gives you the average rate of change because unless you were driving in cruise control the entire time, you definitely were not moving exactly 50 m.p.h. the entire 3 hours.

To find the average rate of change, all you need to do is divide the distance by the time.

Instantaneous Rate of Change. The average rate of change is needed sometimes. However, most of the time what is needed is the rate of change at a certain instant. In your text there is a great example of someone being stopped by a patrol officer for speeding. The officer didn't care that the person was going slowly for a while which "averaged out" the fact that at this particular instant the person was going too fast. Although, it was pretty interesting logic!

So, the question is: How can you find the slope of a line tangent at a certain point of a graph when you have only one point? You actually have all the tools with which to do this now.

Using your graph of $y = x^2$, let's try to look for the rate of change (slope) at a given point, say at $x = 2$. You can draw a secant line between the two points: $(2, 4)$ and $(3, 9)$.

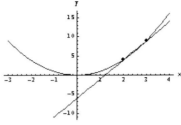

Secant Line

This gives an estimate of the slope of the tangent line at the point $(2,4)$, but not a very good estimate. It would be much better if you chose the second point closer to $x = 2$, and then even closer and then even closer still. Remember, this is what we called a limit.

We are going to let h stand for the distance between $x = 2$ and the second x-coordinate. We want h to get smaller and smaller. In other words, we need the limit as h approaches zero of the slope. Use the slope formula with the two points $(2, f(2))$ and $(2 + h, f(2 + h))$. From this we get one of the most major definitions in calculus, the definition of the derivative which gives us the *slope* of the graph of a function.

$$m_{\tan} = \frac{f(2+h) - f(2)}{(2+h) - 2} = \frac{f(2+h) - f(2)}{h}$$

Example 1: Finding Average Rates of Change from a Table

Suppose the temperature (in °F) at noon in a recent winter month are given by the following table, where t is the tth day of the month in December:

t	0	1	2	3
$f(t)$	70	35	14	7

Find the average rate of change in temperature from $t = 1$ to $t = 2$.

Strategy. Note: The change in t here will correspond to the change in x, which we talked about in the previous problem. Since this is the **average** rate of change, we need to find the slope between the two given points, reading the coordinates off of the chart.

Solution. The average rate of change from $t = 1$ to $t = 2$:

$$\begin{aligned} &= \frac{f(2) - f(1)}{2 - 1} \\ &= \frac{14 - 35}{2 - 1} \\ &= -21\,°\text{F}/\text{d} \end{aligned}$$

Summary. Because the answer is negative, we know the numbers are getting smaller. It is very important to understand **what** we have found. We have **not** found the temperature to be -21 °F on a certain day in December. The answer is -21 °F per day! We have found a rate. We know that the temperature was 21 °F lower (negative sign) on the second day than it was on the first day.

You will be studying rates of change throughout your study of calculus. Be absolutely sure that you completely understand what the answer tells you!

Example 2: Estimating the Slope of a Tangent Line from a Graph

The graph below is the graph of the function $y = \sin x$. Estimate the slope of the tangent line at $x = 0$, $x = \frac{\pi}{2}$, and $x = \pi$.

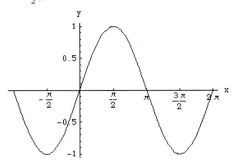

Strategy. Remember that we can easily find the slope of a straight line. We need to draw a straight line through the point in which we are interested, and another point whose coordinates we can estimate relatively easily.

For part a), draw a line tangent to the curve at $x = 0$. Looking at the graph, we can see that this line appears to go through the point $\left(\frac{\pi}{2}, \frac{3}{2}\right)$. We can use this point along with $(0,0)$ to estimate the slope of the graph.

For part b), where $x = \frac{\pi}{2}$, when the tangent line to the graph at the point $\left(\frac{\pi}{2}, 1\right)$ is drawn, the line appears to be horizontal. Make special note of that fact. Horizontal tangent lines will be of great importance as we continue our study of calculus. Also, remember that the slope of a horizontal line is zero!

For part c), where $x = \pi$, we use the same logic as we did in part a using the points $(\pi, 0)$ and $\left(\frac{\pi}{2}, \frac{3}{2}\right)$. From the picture we can tell that this slope is negative and the graph is decreasing. Therefore, when we do the calculations, our answer should be negative.

Solution. a) $x = 0$: We draw the tangent line to $x = 0$, and estimate the slope:

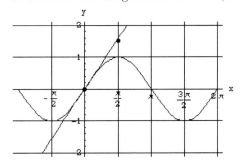

The tangent line is passing through the point $(0,0)$ and approximately through the point $(\pi/2, 3/2)$.

$$\text{Slope} \approx \frac{3/2 - 0}{\pi/2 - 0} = \frac{3}{2} \cdot \frac{2}{\pi} = \frac{3}{\pi}$$

b) $x = \pi/2$: We draw the tangent line to $x = \pi/2$, and estimate the slope:

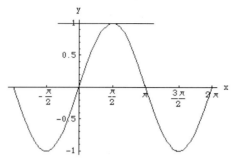

This line appears to be a horizontal line and therefore

$$\text{Slope} \approx 0$$

c) $x = \pi$: We draw the tangent line to $x = \pi$, and estimate the slope:

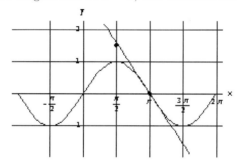

The tangent line is passing through the point $(\pi, 0)$ and approximately through the point $(\pi/2, 3/2)$.

$$\text{Slope} \approx \frac{3/2 - 0}{\pi/2 - \pi} = \frac{3/2}{-\pi/2} = -\frac{3}{2} \cdot \frac{2}{\pi} = -3/\pi$$

Summary. We will use estimation of slopes mostly to be sure that the answer we calculated makes sense. If we sketch a picture and the graph is decreasing at the point of interest, we will check to see that we ended up with a negative answer.

By sketching the tangent line, we have a very basic idea of what type of answer we should get.

2.1. TANGENT LINES AND VELOCITY

Example 3: Finding Instantaneous Rates of Change of a Polynomial

---

Suppose the distance between Bob and a spider t seconds after being spotted is given by $f(t) = 2t^2 + 1$ inches. Find the instantaneous velocity of the spider at time $t = 1$ second.

Strategy. The following problem asks us for instantaneous velocity. Instantaneous velocity is a specific application of instantaneous change. This application is most used in physics. If you are studying business, you will talk about changes in profit, cost, and revenue, in social sciences you would talk about changes in population etc. The word instantaneous tells us that we need the tangent line, and hence the limit. We are to find the instantaneous velocity of the spider at time $t = 1$ second.

Solution. Instantaneous velocity at $t = 1$:

$$
\begin{aligned}
&= \lim_{h \to 0} \frac{f(1+h) - f(1)}{h} &&\text{(of type } \tfrac{0}{0}\text{, manipulate)} \\
&= \lim_{h \to 0} \frac{2(1+h)^2 + 1 - (2 \cdot 1 + 1)}{h} &&\text{(multiply out and cancel)} \\
&= \lim_{h \to 0} \frac{2(1 + 2h + h^2) + 1 - 3}{h} \\
&= \lim_{h \to 0} \frac{2 + 4h + 2h^2 - 2}{h} \\
&= \lim_{h \to 0} \frac{4h + 2h^2}{h} \\
&= \lim_{h \to 0} \frac{h(4 + 2h)}{h} &&\text{(Factor out common } h\text{)} \\
&= \lim_{h \to 0} (4 + 2h) &&\text{(plug in } h = 0\text{)} \\
&= 4
\end{aligned}
$$

Summary. Remember that we have found a **rate**. The correct answer then is **4 inches per second.**

To find instantaneous change, you need to use a lot of algebraic skills. You must compute $f(a + h)$ and $f(a)$. You must subtract correctly. You will ALWAYS end up with h alone in the denominator. Since you cannot have 0 in the denominator, you will algebraically simplify the numerator until the h cancels out.

Example 4: Finding the Equation of a Tangent Line at a Point

Find an equation of the tangent line to $y = \sqrt{2x-1}$ at $x = 5$.

Strategy. All this work has given us the slope of a tangent line. Many times we actually need the equation of this tangent line. Remember, in order to find the equation of a line, we need the slope and one point. We have just found the slope of the line. We were given a point at the beginning of the problem. We use the equation of a line: $y - y_1 = m(x - x_1)$ to start and algebraically manipulate it so that it ends up in the form $y = mx + b$.

Solution. Slope of the tangent line at $x = 5$ is given by

$$
\begin{aligned}
f'(5) &= \lim_{h \to 0} \frac{f(5+h) - f(5)}{h} \quad &&\text{(of type } \tfrac{0}{0}, \text{ manipulate)} \\
&= \lim_{h \to 0} \frac{\sqrt{2(5+h)-1} - \sqrt{2 \cdot 5 - 1}}{h} \\
&= \lim_{h \to 0} \frac{\sqrt{10 + 2h - 1} - \sqrt{9}}{h} \\
&= \lim_{h \to 0} \frac{\sqrt{9 + 2h} - 3}{h} \quad &&\text{(multiply by the conjugate)} \\
&= \lim_{h \to 0} \frac{\sqrt{9 + 2h} - 3}{h} \cdot \frac{\sqrt{9 + 2h} + 3}{\sqrt{9 + 2h} + 3} \\
&= \lim_{h \to 0} \frac{(\sqrt{9 + 2h})^2 - 3^2}{h \cdot (\sqrt{9 + 2h} + 3)} \\
&= \lim_{h \to 0} \frac{9 + 2h - 9}{h \cdot (\sqrt{9 + 2h} + 3)} \\
&= \lim_{h \to 0} \frac{2h}{h \cdot (\sqrt{9 + 2h} + 3)} \\
&= \lim_{h \to 0} \frac{2}{(\sqrt{9 + 2h} + 3)} \\
&= \lim_{h \to 0} \frac{2}{(\sqrt{9} + 3)} = \frac{2}{3+3} = \frac{2}{6} = \frac{1}{3}
\end{aligned}
$$

Now that we know that the slope is $\frac{1}{3}$ and we were given the point $x = 5$, we can get the equation of the line. First of all if $x = 5$, then we can find $y = 3$ from the original equation $y = \sqrt{2x - 1}$. ($y = \sqrt{2(5) - 1} = 3$). Using the equation of a line:

$$
\begin{aligned}
y - 3 &= \frac{1}{3}(x - 5) \\
y - 3 &= \frac{1}{3}x - \frac{5}{3} \\
\implies y &= \frac{1}{3}x + \frac{4}{3}
\end{aligned}
$$

Summary. We have seen the development from average rate of change to instantaneous rate of change which gives us the slope. Then we have seen one problem where we took the slope we found and used it to find the actual equation of the line tangent to the curve at a particular point.

2.2. The Derivative

Key Topics.
- The Derivative at a Point
- The Derivative of a Function as Function
- Differentiable and Non-Differentiable Functions

Worked Examples.
- Finding the Derivative of a Rational Function at a Point
- Finding the Derivative of a Square Root Function
- Sketching the Graph of $f'(x)$ given the Graph of $f(x)$
- Sketching the Graph of $f(x)$ given the Graph of $f'(x)$
- Deciding whether or not a Piece-Wise defined Function is Differentiable at a Point
- Finding Points at which a Function is not Differentiable given the Graph of $f(x)$

Overview

In the previous section we found the slope of the tangent line at a point on the graph of a curve. We will continue to do that, expanding the types of functions in this section. We will go from finding the general derivative to the derivative at a specific point. We are still working from the limit definition of the derivative.

The Derivative at a Point. Here we will continue with what we learned in section 2.1. The only difference is that now the text is actually calling the slope of the tangent line a **derivative**.

The Derivative of a Function as Function. In this section we will be more general. We will find the function which generates the derivative (think slope) at any point, x, on the graph. From that function or formula we can easily find the slope at any specific point on the graph of the function.

Differentiable and Non-Differentiable Functions. As important as it is to know how to find a derivative, it is equally important to know that not all functions have derivatives at all points for which they are defined. Here you will see how to tell quite easily by looking at the graph whether or not the function is differentiable at any given point on the graph.

Example 1: Finding the Derivative of a Rational Function at a Point

Compute the derivative of $f(x) = \frac{2}{x+5}$ at $x = 1$.

Strategy. We will work from the limit definition of derivative at a point here letting $a = 1$. The part of this problem causing trouble is the algebraic simplification. We need to use a common denominator in order to subtract the two fractions. We do it exactly the same way as when subtracting fractions in arithmetic. Try not to let the algebra confuse you.

Solution.

$$
\begin{aligned}
f'(1) &= \lim_{h \to 0} \frac{f(1+h) - f(1)}{h} \\
&= \lim_{h \to 0} \frac{\frac{2}{(1+h)+5} - \frac{2}{1+5}}{h} \\
&= \lim_{h \to 0} \frac{\frac{2}{6+h} - \frac{2}{6}}{\frac{h}{1}} \quad \text{(invert denominator)} \\
&= \lim_{h \to 0} \frac{1}{h} \cdot \left(\frac{2}{6+h} - \frac{1}{3}\right) \quad \text{(use common denominator)} \\
&= \lim_{h \to 0} \frac{1}{h} \cdot \frac{2 \cdot 3 - (6+h)}{3(6+h)} \\
&= \lim_{h \to 0} \frac{1}{h} \cdot \frac{6 - 6 - h}{3(6+h)} \\
&= \lim_{h \to 0} \frac{1}{h} \cdot \frac{-h}{3(6+h)} \quad \text{(cancel common } h\text{'s)} \\
&= \lim_{h \to 0} \frac{-1}{3(6+h)} \quad \text{(substitute 0 for } h\text{)} \\
&= \frac{-1}{3(6+0)} \\
&= -\frac{1}{18}
\end{aligned}
$$

Summary. The answer of $\frac{-1}{18}$ tells us that at the point on the graph where $x = 1$, the slope of the graph is a negative number which lets us conclude that the graph is decreasing at the point $(1, \frac{1}{3})$. The y-coordinate of $\frac{1}{3}$ comes from $f(1) = \frac{2}{1+5} = \frac{2}{6} = \frac{1}{3}$.

Example 2: Finding the Derivative of a Square Root Function

If $f(x) = \sqrt{2x-1}$, find $f'(x)$. Then evaluate the derivative at $x = 1$.

Strategy. In this problem we are asked first for $f'(x)$ and then for $f'(1)$. Actually we will usually be finding the general derivative first and then calculating it at a specific point. All you need to do is substitute x for a in the derivative definition. Therefore, when you finish your calculations, your answer will not be numerical as it was before, but will still contain the letter x. In other words, your derivative will also be a function of x, and hence the notation $f'(x)$.

Solution.

$$\begin{aligned}
f'(x) &= \lim_{h \to 0} \frac{f(x+h) - f(x)}{h} \\
&= \lim_{h \to 0} \frac{\sqrt{2(x+h) - 1} - \sqrt{2x-1}}{h} & \text{(multiply by conjugate)} \\
&= \lim_{h \to 0} \frac{(\sqrt{2(x+h) - 1} - \sqrt{2x-1})}{h} \cdot \frac{(\sqrt{2(x+h) - 1} + \sqrt{2x-1})}{(\sqrt{2(x+h) - 1} + \sqrt{2x-1})} \\
& \qquad\qquad \text{(multiply the binomials in the numerator together)} \\
&= \lim_{h \to 0} \frac{(\sqrt{2(x+h) - 1})^2 - (\sqrt{2x-1})^2}{h \cdot (\sqrt{2(x+h) - 1} + \sqrt{2x-1})} & \text{(simplify numerator)} \\
&= \lim_{h \to 0} \frac{2(x+h) - 1 - (2x - 1)}{h \cdot (\sqrt{2(x+h) - 1} + \sqrt{2x-1})} \\
&= \lim_{h \to 0} \frac{2x + 2h - 1 - 2x + 1}{h \cdot (\sqrt{2(x+h) - 1} + \sqrt{2x-1})} \\
&= \lim_{h \to 0} \frac{2h}{h \cdot (\sqrt{2(x+h) - 1} + \sqrt{2x-1})} & \text{(cancel common } h\text{'s)} \\
&= \lim_{h \to 0} \frac{2}{\sqrt{2(x+h) - 1} + \sqrt{2x-1}} & \text{(substitute 0 for } h\text{)} \\
&= \frac{2}{\sqrt{2x-1} + \sqrt{2x-1}} \\
&= \frac{2}{2\sqrt{2x-1}} \\
&= \frac{1}{\sqrt{2x-1}}
\end{aligned}$$

And the derivative at $x = 1$ is given by $f'(1) = \frac{1}{\sqrt{2 \cdot 1 - 1}} = \frac{1}{\sqrt{1}} = 1$.

Summary. We got the answer of 1 when we calculated $f'(1)$ which tells us that at the point in the graph where $x = 1$, the slope of the graph is positive 1 and therefore the graph is increasing.

Example 3: Sketching the Graph of $f'(x)$ given the Graph of $f(x)$

Given the graph of $f(x)$ below, sketch a plausible graph of $f'(x)$.

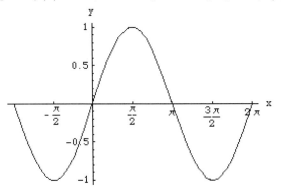

Strategy. To begin this problem look at the points where the tangent line is horizontal. Since the slope of a horizontal line is zero, these points will give the x-intercepts on the graph of the derivative. For this particular graph these points are $(-\frac{\pi}{2}, 0)$, $(\frac{\pi}{2}, 0)$, and $(\frac{3\pi}{2}, 0)$. Then we need to estimate slopes for some x-values in between the ones already graphed. Go through the solution for hints on how to do that.

Solution. The given graph is actually the same graph that we saw in the previous section, example 3. In example 3, we estimated the slope of the tangent line at $x = 0$, $x = \frac{\pi}{2}$, and $x = \pi$.

Slope at $x = 0$ is approximately $\frac{3}{\pi}$ and since $\pi \approx 3$, slope $\approx \frac{3}{3} = 1$

Slope at $x = \pi/2$ is approximately 0

Slope at $x = \pi$ is approximately $-3/\pi \approx -1$

Following this pattern, the function values of f' are approximately given by the following table:

x	$-\pi$	$-\pi/2$	0	$\pi/2$	π	$3\pi/2$	2π
$f'(x)$	-1	0	1	0	-1	0	1

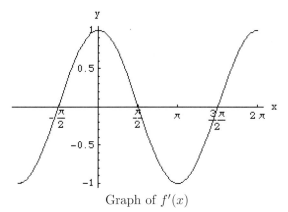
Graph of $f'(x)$

The graph resembles $y = \cos x$. We will learn later that $f'(x) = \cos x$ for $f(x) = \sin x$.

Summary. It takes some time to learn and understand how to graph the derivative of the function. When you are able to do that, you will better understand the connection between slopes, derivatives and rates of change.

Example 4: Sketching the Graph of $f(x)$ given the Graph of $f'(x)$

Given the graph of $f'(x)$ below, sketch a plausible graph of $f(x)$ with $f(0) = 0$.

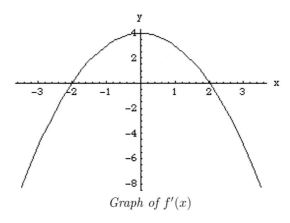

Graph of $f'(x)$

Strategy. Since this is a graph of a derivative, we first look for the places where the graph is above the x-axis. This will tell us where the slope of the original graph is positive and, hence, on what intervals the original function is increasing. Similarly, when the graph of the derivative is below the x-axis, the original function will be decreasing. Finally, the y-coordinate of all the x-intercepts is 0. The slope of the original graph then is 0 at those points. Therefore, there will be horizontal tangents at these points. The *peaks* and *valleys* of the graph occur at those places. Carefully follow the solution to see this strategy actually work!

Solution. $f'(x)$ is negative for $x < -2$ and $x > 2$, implying that the graph of f is decreasing on these intervals as we go from left to right.

$f'(x)$ is positive for $-2 < x < 2$, implying that the graph of f is increasing on the interval $(-2, 2)$.

At $x = -2$ the graph has slope 0, the slope goes from negative to positive as we go from left to right, therefore at $x = -2$ the graph of f has a low point.

At $x = 2$ the graph has slope 0, the slope goes from positive to negative, therefore at $x = 2$ the graph of f has a high point.

A possible graph:

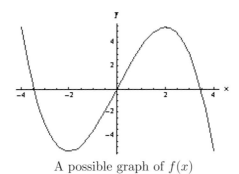

A possible graph of $f(x)$

Summary. As was said with the previous problem, it takes time to understand how to go from one graph to the other. After you have done several, it will make more sense and your understanding of the derivative will grow.

Example 5: Deciding whether or not a Piece-Wise defined Function is Differentiable at a Point

---

Is the function $f(x)$ given below continuous and differentiable at $x = 0$, and $x = 1$?

$$f(x) = \begin{cases} 2x^2 & \text{if} \quad x \leq 0 \\ 4x & \text{if} \quad 0 < x < 1 \\ 3 & \text{if} \quad x \geq 1 \end{cases}$$

Strategy. We will look for continuity by seeing if the limit from the left is equal to the limit from the right, and also if the function value is the same as those limits. If all three are the same, then we can conclude that the function is continuous at the given point.

We will look for differentiability by using the definition of the derivative, and seeing whether the right and left hand limits are the same. If they are, then we can conclude that the function is differentiable at the given point.

It will also help a lot if you sketch a graph and see what this function looks like.

Solution. a) $x = 0$

First we check to see if f is continuous at $x = 0$.

$$\lim_{x \to 0^-} f(x) = \lim_{x \to 0} 2x^2 = 2 \cdot 0^2 = 0$$

$$\lim_{x \to 0^+} f(x) = \lim_{x \to 0} 4x = 4 \cdot 0 = 0$$

Since both limits are the same, it follows that $\lim_{x \to 0} f(x) = 0$. Also note that $f(0) = 0$. Hence, f is continuous at $x = 0$.

Next we check to see if f is differentiable at $x = 0$.
Left-Hand Derivative

$$\begin{aligned} D_-f(0) &= \lim_{h \to 0^-} \frac{f(0+h) - f(0)}{h} \\ &= \lim_{h \to 0^-} \frac{2(0+h)^2 - 2(0)^2}{h} \\ &= \lim_{h \to 0^-} \frac{2h^2 - 0}{h} \\ &= \lim_{h \to 0^-} 2h \\ &= 0 \end{aligned}$$

Right-Hand Derivative

$$\begin{aligned} D_+f(0) &= \lim_{h \to 0^+} \frac{f(0+h) - f(0)}{h} \\ &= \lim_{h \to 0^+} \frac{4(0+h) - 4(0)}{h} \\ &= \lim_{h \to 0^-} \frac{4h - 0}{h} \end{aligned}$$

$$= \lim_{h \to 0^-} 4$$
$$= 4$$

Since the Right and Left-Hand Limits are different, f is not differentiable at $x = 0$.

b) $x = 1$
First we check to see if f is continuous at $x = 1$.

$$\lim_{x \to 1^-} f(x) = \lim_{x \to 1} 4x = 4 \cdot 1 = 4$$
$$\lim_{x \to 1^+} f(x) = \lim_{x \to 1} 3 = 3$$

Since the two limits are not equal it follows that f is not continuous at $x = 1$. Because the function is not continuous at this point it follows that f cannot be differentiable at $x = 1$.

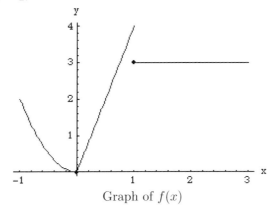

Graph of $f(x)$

Summary. Hopefully, the previous problem helped you see the connection between the graph of a function, the continuity of a function, and the differentiability of a function.

Example 6: Finding Points at which a Function is not Differentiable given the Graph of $f(x)$

Find all points at which the function is not differentiable.

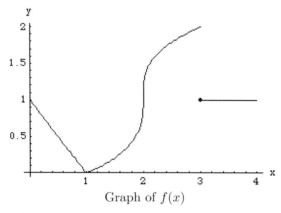

Graph of $f(x)$

Strategy. The key thing to remember when looking at a graph for points where the graph is not differentiable, is that most of the time most of the graphs are differentiable. So we are looking for the exceptions. These exceptions occur when the graph is discontinuous in any way, has a vertical slope, or comes to a *pointy* corner.

Solution. f is not differential at $x = 1$, because f has a corner point here,
f is not differential at $x = 2$, because f has a vertical tangent line at this point,
and f is not differential at $x = 3$, because f has a jump at $x = 3$.

Summary. In the last two problems, we have encountered how to determine points of continuity and differentiability both graphically and algebraically. In the next sections we will learn several formulas which will make the finding of the derivatives a lot faster and a lot easier!

2.3. The Power Rule

Key Topics.
- Power Rule
- General Derivative Rules
- Higher Derivatives

Worked Examples.
- Finding the Derivative of a Sum
- Rewriting a Function before finding the Derivative
- Finding the Equation of a Tangent Line
- Computing Higher Derivatives

Overview

We have finally come to the part of the course where we can use formulas instead of the limit definition to find the derivative. It really helps if you learn to say the formulas in words.

Power Rule. The power rule says that the derivative of a constant is 0 and that the derivative of x is 1. The main part and the very important part of the power rule says to **bring the power in front, repeat the function and decrease the power by** 1. The "repeat the function" part will be very important later on when the functions become more complex.

General Derivative Rules. In this section we will see that taking the derivative of a sum or difference of terms is done exactly the way we intuitively want to do it. We get to take the derivative of one term at a time.

Higher Derivatives. The 2nd derivative is simply the derivative of the 1st derivative. The 3rd derivative is the derivative of the 2nd derivative and so on.

Example 1: Finding the Derivative of a Sum

If $f(x) = 3x^2 - \frac{2}{x} + \sqrt[3]{x}$, find $f'(x)$.

Strategy. Since the only rule we have is the power rule, we must change all the terms so that they contain x to a power rather than having x in the denominator or under a radical. In this problem $\frac{-2}{x}$ is rewritten as $-2x^{-1}$ and $\sqrt[3]{x}$ is rewritten as $x^{\frac{1}{3}}$.

Solution. First rewrite $f(x)$:

$$f(x) = 3x^2 - 2x^{-1} + x^{\frac{1}{3}}$$

Then using the power and general derivative rule, we have

$$\begin{aligned} f'(x) &= 3(2x) - 2(-x^{-2}) + \frac{1}{3}x^{\frac{1}{3}-1} \\ &= 6x + 2x^{-2} + \frac{1}{3}x^{\frac{-2}{3}} \\ &= 6x + \frac{2}{x^2} + \frac{1}{3\sqrt[3]{x^2}} \end{aligned}$$

Summary. If you get confused when changing roots to fractional exponents, think about a tree. The roots grow down! The "down" number in the fractional exponent is the root.

Example 2: Rewriting a Function before finding the Derivative

Find the derivative of
$$f(t) = \frac{2\sqrt{t} - 3\sqrt[3]{t}}{\sqrt{t}}$$

Strategy. We do not yet have any rule for quotients, so once again we need to rewrite the problem so that all the terms are variables with powers and coefficients. Only then can we use the power rule to get the derivative.

Solution. First rewrite $f(t)$:

$$\begin{aligned}
f(t) &= \frac{2\sqrt{t} - 3\sqrt[3]{t}}{\sqrt{t}} \\
&= \frac{2\sqrt{t}}{\sqrt{t}} - \frac{3\sqrt[3]{t}}{\sqrt{t}} \\
&= 2 - 3\left(\frac{t^{\frac{1}{3}}}{t^{\frac{1}{2}}}\right) \\
&= 2 - 3t^{\frac{1}{3} - \frac{1}{2}} \\
&= 2 - 3t^{\frac{2}{6} - \frac{3}{6}} \\
&= 2 - 3t^{\frac{-1}{6}}
\end{aligned}$$

Then using the power and general derivative rules, we have

$$\begin{aligned}
f'(t) &= 0 - 3\left(\frac{-1}{6} t^{\frac{-1}{6} - 1}\right) \\
&= -3\left(-\frac{1}{6} t^{\frac{-7}{6}}\right) \\
&= \frac{1}{2} t^{\frac{-7}{6}} \\
&= \frac{1}{2\sqrt[6]{t^7}}
\end{aligned}$$

Summary. If you go slowly through the steps, and carefully rewrite everything so that you only have terms with powers before you begin to differentiate, then you will do fine!

Example 3: Finding the Equation of a Tangent Line

If $f(x) = 3x^2 - 5x + 1$, then find an equation of the tangent line at $x = 1$.

Strategy. We did a problem finding the equation of a tangent line in an earlier section. At that time, we used the limit definition to find the slope. From now on we do not need to use the limit definition. We can just use the power rule to get the derivative which gives us the slope. Then we will use our slope and the given point to find the equation of the tangent line.

Solution. $f'(x) = 6x - 10$, making $f'(1) = 6 \cdot 1 - 10 = -4$. The slope of the tangent line is thus -4. A point on the tangent line is given by $(1, f(1))$. Since $f(1) = 3 \cdot 1^2 - 5 \cdot 1 + 1 = -1$, the needed point is $(1, -1)$.

The line with slope -4 through the point $(1, -1)$ has equation

$$\begin{aligned} y - (-1) &= -4(x - 1) \\ y + 1 &= -4x + 4 \\ y &= -4x + 3 \end{aligned}$$

Summary. Now that we have the power rule, finding the derivative, and hence the slope, is a much easier and quicker process. If necessary, review the algebra needed to find the equation of a line, as you will be using it a lot in your study of calculus.

Example 4: Computing Higher Derivatives

If $f(x) = 3x^2 - 5$, find the second derivative $f''(x)$.

Strategy. To find the 2nd derivative we need to take the derivative of the 1st derivative. So we will go through the process twice.

Solution.
$$\begin{aligned} f'(x) &= 3(2x) - 5 \cdot 0 = 6x \\ f''(x) &= 6 \end{aligned}$$

Summary. Once you understand how to use the rules for differentiation, the higher order derivatives should not pose much of a problem. You just keep repeating the process over and over until you get to the level you need.

2.4. Product and Quotient Rule

Key Topics.
- The Product Rule
- The Quotient Rule

Worked Examples.
- Using the Product Rule
- Using the Quotient Rule
- Using the Product and Quotient Rule
- Another Quotient Rule Example

Overview

In the last section, we found that to take the derivative of the sum or difference of terms that we were able to take the derivative of each term separately. This is what we intuitively want to do. Unfortunately, the product and quotient rules are not nearly so nice. We will give you ways to say them in words which should help you to remember them.

The Product Rule. You can think of the product rule in terms of the functions $u(x)$ and $v(x)$ or you can think of the two functions as the first function and the second function. In that case the product rule says to take the second function times the derivative of the first plus the first times the derivative of the second. A lot of people find the rule easier to remember that way!

The Quotient Rule. In a similar fashion to the way you can remember the product rule, you can say the quotient rule as *bottom times derivative of top minus top times derivative of bottom all over bottom squared*. It looks like:

$$\frac{\text{bottom times derivative of top - top times derivative of bottom}}{(\text{bottom})^2}$$

Another fun way to remember it is:

$$\frac{\text{lo di hi - hi di lo}}{\text{lo lo}}$$

Because subtraction is not commutative, it is really important that you subtract in correct order!

Example 1: Using the Product Rule

If $f(x) = (2\sqrt{x} - 3x^3)(4x - 10)$, use the product rule to find $f'(x)$.

Strategy. In this problem the first function is $u(x) = 2\sqrt{x} - 3x^3$ and the second function is $v(x) = 4x - 10$. We can use the chart below to find both of the derivatives before we put it all together.

Solution. $f(x)$ is a product of two functions $u(x) = 2\sqrt{x} - 3x^3$, and $v(x) = 4x - 10$. Before using the product rule, we find the derivative of the function u.

$$\begin{aligned} u(x) &= 2\sqrt{x} - 3x^3 = 2x^{\frac{1}{2}} - 3x^3 \\ u'(x) &= 2(\frac{1}{2}x^{-\frac{1}{2}}) - 3(3x^2) = x^{\frac{-1}{2}} - 9x^2 = \frac{1}{\sqrt{x}} - 9x^2 \end{aligned}$$

We have,

$$\begin{array}{ll} u(x) = 2\sqrt{x} - 3x^3 & v(x) = 4x - 10 \\ u'(x) = \frac{1}{\sqrt{x}} - 9x^2 & v'(x) = 4 \end{array}$$

Using the product rule,

$$\begin{aligned} f'(x) &= u'(x) \cdot v(x) + u(x) \cdot v'(x) \\ &= (\frac{1}{\sqrt{x}} - 9x^2)(4x - 10) + (2\sqrt{x} - 3x^3)(4) \end{aligned}$$

Do we need to simplify this equation? If we are looking for a simplified answer then it would have been easier to expand $f(x)$ before taking the derivative. Only simplify if you are asked to simplify, or if you have to find higher derivatives.

Summary. Before you move on to the quotient rule, it would be a good idea to work several problems with the product rule. If you try to learn both of them at the same time, you may get them confused. So make sure you completely understand the product rule before you attempt the quotient rule.

Example 2: Using the Quotient Rule

If $f(x) = \dfrac{3x^4 - 2x^2 + 10}{2x - 9}$, find $f'(x)$.

Strategy. In this problem the top function is $u(x) = 3x^4 - 2x^2 + 10$, and the bottom function is $v(x) = 2x - 9$. As in the product rule, we can use the chart below to find all the needed derivatives before we put it all together.

Solution. $f(x)$ is a quotient of two functions $u(x) = 3x^4 - 2x^2 + 10$, and $v(x) = 2x - 9$.

$$u(x) = 3x^4 - 2x^2 + 10 \quad v(x) = 2x - 9$$
$$u'(x) = 12x^3 - 4x \quad\quad\quad v'(x) = 2$$

$$\begin{aligned}
f'(x) &= \frac{u' \cdot v - u \cdot v'}{v^2} \\
&= \frac{(12x^3 - 4x)(2x - 9) - (3x^4 - 2x^2 + 10)(2)}{(2x - 9)^2}
\end{aligned}$$

(multiply factors in the numerator only, and combine like terms)

$$\begin{aligned}
&= \frac{24x^4 - 108x^3 - 8x^2 + 36x - 6x^4 + 4x^2 - 20}{(2x - 9)^2} \\
&= \frac{18x^4 - 108x^3 - 4x^2 + 36x - 20}{(2x - 9)^2}
\end{aligned}$$

Summary. One of the most common mistakes is to set up the numerator "backwards". Be very careful of that since it will give you the opposite answer. Whichever way you choose to memorize the formula, just remember the bottom goes first!

Example 3: Using the Product and Quotient Rule

Find the derivative of the function
$$f(x) = \frac{(2\sqrt{x} - 10)(\frac{2}{x} + 5x)}{x^2 + 1}$$

Strategy. In this problem we have both a product and a quotient. One way to work it is to multiply out the numerator and then use the quotient rule. Another way is to work the top as a product and also use the quotient rule. It is good to see a problem where both is used. In this problem we have a choice, in later problems we will have to use both rules in the same problem.

Solution. This function is a product and a quotient of functions. Think of f as a quotient $f(x) = \dfrac{u(x)}{v(x)}$.

First we use the product rule to find the derivative of the function
$$u(x) = (2\sqrt{x} - 10)(\frac{2}{x} + 5x)$$

$$\begin{aligned} u'(x) &= \frac{d}{dx}(2\sqrt{x} - 10)\left(\frac{2}{x} + 5x\right) + (2\sqrt{x} - 10)\frac{d}{dx}\left(\frac{2}{x} + 5x\right) \\ &= \left(\frac{1}{\sqrt{x}}\right)\left(\frac{2}{x} + 5x\right) + (2\sqrt{x} - 10)\left(-\frac{2}{x^2} + 5\right) \end{aligned}$$

In this example we will not simplify this expression.

$$\begin{aligned} f'(x) &= \frac{u' \cdot v - u \cdot v'}{v^2} \\ &= \frac{\left[\left(\frac{1}{\sqrt{x}}\right)\left(\frac{2}{x} + 5x\right) + (2\sqrt{x} - 10)\left(-\frac{2}{x^2} + 5\right)\right](x^2 + 1) - \left[(2\sqrt{x} - 10)\left(\frac{2}{x} + 5x\right)\right](2x)}{(x^2 + 1)^2} \end{aligned}$$

Summary. There are a lot of terms in this answer. Once again, if you work slowly and pay close attention to the small details, you will arrive at the correct answer.

Example 4: Another Quotient Rule Example

Suppose all we know about a function is the following:

$$f(x) = \frac{g(x)}{2x^2 - 1},\ g(2) = 3,\ g'(2) = 5$$

Find $f'(2)$.

Strategy. In this problem we have $g(x)$, an unknown function, on the top. Even though we do not know the exact function, we have enough information given to us to find the derivative of $f(x)$ evaluated at $x = 2$.

Solution.

$$f(x) = \frac{g(x)}{v(x)}$$

$$f'(x) = \frac{g'(x) \cdot v(x) - g(x) \cdot v'(x)}{[v(x)]^2}$$

$$f'(2) = \frac{g'(2) \cdot v(2) - g(2) \cdot v'(2)}{[v(2)]^2}$$

We are given: $g(2) = 3$, and $g'(2) = 5$

We know that $v(x) = 2x^2 - 1$ making $v(2) = 7$, and $v'(x) = 4x$ making $v'(2) = 8$. Placing all the values in their correct places, we have:

$$f'(2) = \frac{5 \cdot 7 - 3 \cdot 8}{7^2} = \frac{11}{49}$$

Summary. Now we have seen derivatives of the sum, difference, product, and quotients of algebraic functions. In the next section we will greatly expand the functions we can differentiate as we learn the chain rule.

2.5. The Chain Rule

Key Topics.
- The Chain Rule
- Multiple Chain Rules

Worked Examples.
- Chain Rule on a Square-Root Function
- Another Chain Rule Example
- A Derivative involving Multiple Chain Rules

Overview

The chain rule allows us to find the derivatives of much more complex functions. Before this section, we were easily able to find the derivative of $f(x) = x^4$, but not of $g(x) = (10x^2 + 5)^4$. Now we will be able to do that also.

The Chain Rule. This is called the chain rule because the differentiation is done as a chain of events. If we have functions inside of functions (i.e.: composition of functions), we need a logical method to make sure that all the derivatives are taken correctly.

Multiple Chain Rules. If the function is even more complex, we may have to use the chain rule repeatedly.

Example 1: Chain Rule on a Square-Root Function

If $f(x) = \sqrt{2x^2 + 10}$, find $f'(x)$.

Strategy. The very first thing to do here is to change the radical to a fractional exponent. Then the function becomes: $f(x) = (2x^2 + 10)^{\frac{1}{2}}$. The *major* thing that is happening here is that a function is being raised to the one-half power. That is what we call the outer function. In addition to that, we have the function which is being raised to the one-half power - that is what we call the inner function. In words, we need to bring the power ($\frac{1}{2}$) to the front, **repeat the function** down a power, and then multiply by the derivative of the inner function.

Solution. The inner function is $u = 2x^2 + 10$, and the outer function is $y = \sqrt{u} = u^{\frac{1}{2}}$.

The derivative of the inner function: $\dfrac{du}{dx} = 4x$

The derivative of the outer function: $\dfrac{dy}{du} = \dfrac{1}{2}u^{\frac{-1}{2}} = \dfrac{1}{2\sqrt{u}} = \dfrac{1}{2\sqrt{2x^2 + 10}}$

$$\begin{aligned} f'(x) &= \text{(derivative of outer function)} \cdot \text{(derivative of inner function)} \\ &= \dfrac{1}{2\sqrt{2x^2 + 10}} \cdot 4x = \dfrac{4x}{2\sqrt{2x^2 + 10}} = \dfrac{2x}{\sqrt{2x^2 + 10}} \end{aligned}$$

Summary. As stated in the strategy section above, one way to think about this is to remember that the power comes in front, then repeat the function down one degree, and then multiply by the derivative of the inner function. That would look like: $f'(x) = \frac{1}{2}(2x^2 + 10)^{\frac{1}{2} - 1}(4x)$ which simplifies to $\frac{2x}{\sqrt{2x^2+10}}$ which is, of course, the same answer we got writing out all the different parts separately. Both ways are great - decide which way is easier for you to understand!

Example 2: Another Chain Rule Example

Find:
$$\frac{d}{dx}\left(\frac{2}{(10x^3 - 10)^2}\right)$$

Strategy. Start this problem by rewriting the function so that there is not a quotient. When the numerator is a constant it is usually easier to rewrite the function and then use the power rule instead just using the quotient rule without making any changes first.

Solution. The inner function is $u = 10x^3 - 10$, and the outer function is:
$$y = \frac{2}{u^2} = 2u^{-2}$$

$$\text{The derivative of the inner function } \frac{du}{dx} = 30x^2$$

$$\begin{aligned}
\text{The derivative of the outer function is } \frac{dy}{du} &= \frac{d}{du}\left(2u^{-2}\right) \\
&= 2(-2u^{-3}) = \frac{-4}{u^3} \\
&= \frac{-4}{(10x^3 - 10)^3}
\end{aligned}$$

$$\begin{aligned}
\frac{d}{dx}\left(\frac{2}{(10x^3 - 10)^2}\right) &= (\text{derivative of outer function}) \cdot (\text{derivative of inner function}) \\
&= \frac{-4}{(10x^3 - 10)^3} \cdot 30x^2 \\
&= \frac{-120x^2}{(10x^3 - 10)^3}
\end{aligned}$$

Summary. As in the previous problem, we can rewrite the original problem and then find a "chain of derivatives":

$$\begin{aligned}
\frac{d}{dx}\left(\frac{2}{(10x^3 - 10)^2}\right) &= \frac{d}{dx}\left(2(10x^3 - 10)^{-2}\right) \\
&= -4(10x^3 - 10)^{-3})30x^2 \\
&= \frac{-120x^2}{(10x^3 - 10)^3}
\end{aligned}$$

Example 3: A Derivative involving Multiple Chain Rules

Compute the derivative of $f(x) = \left(\sqrt{2x+1}+2\right)^3$.

Strategy. Once again, the first thing to do is change the root to a fractional exponent and then find the derivative of the outer function and multiply it by the derivative of the inner function.

Solution. The inner function is $u = \sqrt{2x+1}+2 = (2x+1)^{\frac{1}{2}}+2$, and the outer function is $y = u^3$. Note that the derivative of the inner function also requires the chain rule.

$$\text{The derivative of the inner function is } \begin{aligned} \frac{du}{dx} &= \frac{1}{2}(2x+1)^{\frac{-1}{2}} \cdot \frac{d}{dx}(2x+1) \\ &= \frac{1}{2\sqrt{2x+1}} \cdot 2 \\ &= \frac{1}{\sqrt{2x+1}} \end{aligned}$$

$$\text{The derivative of the outer function is } \frac{dy}{du} = 3u^2 = 3\left(\sqrt{2x+1}+2\right)^2$$

$$\begin{aligned} f'(x) &= (\text{derivative of outer function}) \cdot (\text{derivative of inner function}) \\ &= 3\left(\sqrt{2x+1}+2\right)^2 \cdot \frac{1}{\sqrt{2x+1}} \\ &= \frac{3\left(\sqrt{2x+1}+2\right)^2}{\sqrt{2x+1}} \end{aligned}$$

Summary. If you want to think of the process more as a chain you can take a *chain of derivatives.*

$$\begin{aligned} \frac{d}{dx}\left((2x+1)^{\frac{1}{2}}+2\right)^3 &= 3\left((2x+1)^{\frac{1}{2}}+2)^2\right) \cdot \frac{1}{2}(2x+1)^{\frac{-1}{2}} \cdot 2 \\ &= \frac{3((2x+1)^{\frac{1}{2}}+2)^2}{\sqrt{2x+1}} \end{aligned}$$

Once again, whichever method you find easier is the one for you to use.

2.6. Derivatives of Trigonometric Functions

Key Topics.
- The Derivative of Sine and Cosine
- The Derivatives of other Trigonometric Functions

Worked Examples.
- A Derivative that Requires the Chain Rule
- A Derivative that Requires the Product and the Chain Rule
- Derivative of $\sec x$
- Another Derivative Requiring the Chain Rule

Overview

We are now going to use the derivatives of a new class of functions - the trigonometric functions. Once we know the derivatives of the sine and cosine function, the other 4 can easily be found by rewriting them in terms of sine and cosine and then using the quotient rule. One major thing to remember is that the derivative of all of the trig functions beginning with "co" (cosine, cosecant, and cotangent) are negative. The other three derivatives are positive!

The Derivative of Sine and Cosine. Just as we would like, the derivative of the sine function is the cosine function. The derivative of the cosine function is negative the sine function. Written in symbols:

$$\frac{d}{d\theta} \sin \theta = \cos \theta$$
$$\frac{d}{d\theta} \cos \theta = -\sin \theta$$

As stated above, only the derivative of the cosine function is negative.

The Derivatives of other Trigonometric Functions. By rewriting the other trig formulas in terms of sine and cosine, we can easily find their derivatives using the quotient rule. In example 3 we find the derivative of the secant function in this manner. The derivatives are as follows:

$$\frac{d}{d\theta} \tan \theta = \sec^2 \theta$$
$$\frac{d}{d\theta} \cot \theta = -\csc^2 \theta$$
$$\frac{d}{d\theta} \sec \theta = \sec \theta \tan \theta$$
$$\frac{d}{d\theta} \csc \theta = -\csc \theta \cot \theta$$

You need to learn these. Notice which derivatives are negative and which are positive. Also, notice that the secant and tangent are always together as are the cosecant and the cotangent. These little hints should help you remember them.

2.6. DERIVATIVES OF TRIGONOMETRIC FUNCTIONS

Example 1: A Derivative that Requires the Chain Rule

If $f(x) = \cos(2x + 1)$, find $f'(x)$.

Strategy. When differentiating a trig function remember the letters PTA - yes, just like the group your parents went to when you were in grade school. First we deal with the Power, then the Trig function and lastly, the Angle. One step at a time.

Solution. P: there is no power here so we don't have to worry about it.
T: the derivative of $\cos\theta$ is $-\sin\theta$. Note: the angle matches!
A: the derivative of the angle here is 2

Putting this all together: $\frac{d}{dx}\cos(2x+1) = -\sin(2x+1) \cdot 2 = -2\sin(2x+1)$

Summary. If you go one step at a time and remember PTA, then the trig function derivatives will come easily to you.

Example 2: A Derivative that Requires the Product and the Chain Rule

If $f(t) = \sin(5t)\cos(3t)$, find $f'(t)$.

Strategy. Here we have both a product and a chain rule. The first function is $\sin(5t)$. The second function is $\cos(3t)$. For both functions we will use PTA, and in both cases we don't have a power to differentiate.

Solution. We use the product rule.
$$u(t) = \sin(5t) \qquad v(t) = \cos(3t)$$
$$u'(t) = 5\cos(5t) \qquad v'(t) = -3\sin(3t)$$

$$\begin{aligned} f'(t) &= u'v + uv' \\ &= (5\cos(5t))(\cos(3t)) + (\sin(5t))(-3\sin(3t)) \\ &= 5\cos(5t)\cos(3t) - 3\sin(5t)\sin(3t) \end{aligned}$$

Summary. Are you starting to understand how to differentiate these trig functions? The next example has the secant function.

Example 3: Derivative of $\sec x$

Compute the derivative of $f(x) = \sec x$, and then find the derivative of $f(t) = \sec(5t^2 - 4)$.

Strategy. To compute the derivative of $\sec x$, we need to rewrite it as $\dfrac{1}{\cos x}$ and then use the quotient rule. After we do that, we will find the derivative of $f(t) = \sec(5t^s - 4)$ using PTA as we did in the previous two examples.

Solution.
$$f(x) = \sec x = \frac{1}{\cos x}$$
We can use the quotient rule to find the derivative.

$$\begin{array}{ll} u(x) = 1 & v(x) = \cos x \\ u'(t) = 0 & v'(t) = -\sin x \end{array}$$

$$\begin{aligned} f'(t) &= \frac{u'v - uv'}{v^2} \\ &= \frac{0 \cdot \cos x - 1 \cdot (-\sin x)}{(\cos x)^2} \\ &= \frac{\sin x}{\cos^2 x} \\ &= \frac{1}{\cos x} \cdot \frac{\sin x}{\cos x} \\ &= \sec x \tan x \end{aligned}$$

In the second part of this problem, in order to find the derivative of $f(t) = \sec(5t^2 - 4)$, we start with the derivative of the secant function (which is the outer function) and then take the derivative of the angle (which is the inner function).

$$\begin{aligned} \frac{d}{dt} \sec(5t^2 - 4) & \\ = \sec(5t^2 - 4) \tan(5t^2 - 4) \cdot 10t & \\ = 10t \sec(5t^2 - 4) \tan(5t^2 - 4) & \end{aligned}$$

Summary. Very important - when you differentiate the trig function part, you **keep the angle the same**. Then you multiply by the derivative of the angle. An example of an extremely common mistake is to write the derivative of $y = \sin(x^2)$ as $\cos(2x)$. The derivative of the trig function gets the same angle as the original function. The derivative of the angle comes next, thus $\left(\sin(x^2)\right)' = \cos(x^2) \cdot 2x \neq \sin(2x)$.

Example 4: Another Derivative Requiring the Chain Rule

If $f(x) = \cos^3(5x + x^2)$, find $f'(x)$.

Strategy. Here we get to use all of PTA. We start by differentiating the 3rd power - remember to differentiate a power, we bring the power in front and **repeat the same function** raised to one power less. Next we differentiate the cosine function keeping the angle the same. Lastly, we differentiate the angle.

Solution. P: the 3 will come in front and the cosine function will be raised to a power of 2
T: the derivative of the cosine function is negative the sine of the **same** angle
A: lastly we differentiate the angle

$$\begin{aligned}
&\frac{d}{dx}\cos^3(5x + x^2) \\
= \ & (3\cos^2(5x + x^2))(-\sin(5x + x^2))(5 + 2x) \\
= \ & -(15 + 6x)\cos^2(5x + x^2)\sin(5x + x^2)
\end{aligned}$$

Summary. It really helps if you separate the trigonometric derivatives into the three parts and make sure you work each part separately. Then you won't be combining steps that are not supposed to be combined.

2.7. Derivatives of Exponential and Logarithmic Functions

Key Topics.
- Derivatives of Exponential Functions
- Derivatives of Logarithmic Functions
- Logarithmic Differentiation

Worked Examples.
- Derivatives of a Sum of Simple Exponential and Logarithmic Functions
- Chain Rule on an Exponential Function
- Chain Rule on a Logarithmic Function
- Two Examples of Logarithmic Differentiation

Overview

Now we come to another class of functions - the transcendental functions. These include logarithmic and exponential functions.

An exponential function is a function where the variable appears in the exponent. Some examples of exponential functions are $e^x, 2^{x^2}, 3^{x+7}$, etc. (x^2, x^3, and $x^{\frac{1}{2}}$ are **not** exponential functions. They are algebraic and their derivatives follow the power rule.)

Logarithmic functions have a logarithm in them.

Derivatives of Exponential Functions. The derivatives of exponential functions work in an interesting way. If $f(x) = e^x$, then $f'(x) = e^x$. That's right, the function and the derivative are exactly the same. This is the only function where this is true.

If the power is a function of x, then we need the chain rule also. To differentiate e to any power,

1) rewrite exactly what is given and
2) multiply by the derivative of the power

For example, if $f(x) = e^{x^3}$, then $f'(x) = e^{x^3} \cdot 3x^2 = 3x^2 e^{x^3}$

Hopefully, this is easily understood. But what happens if the base isn't e? Then we do the exact same process, and multiply by ln of the base.

$$\frac{d}{dx} a^x = a^x \ln a$$

For example, if $g(x) = 2^{x^3}$, then $f'(x) = (2^{x^3})(\ln 2)(3x^2)$.
That's all there is to it!

Derivatives of Logarithmic Functions. The derivative of the logarithmic function is not much harder than the exponential function.

$$\frac{d}{dx}(\ln x) = \frac{1}{x}$$

If necessary, we use the chain rule here also, just like in any other function.
For example, if $f(x) = \ln(x^3 + 2x)$, then $f'(x) = \dfrac{1}{x^3 + 2x}(3x^2 + 2) = \dfrac{3x^2 + 2}{x^3 + 2x}$.

Once again, we also have to deal with bases other than e. If the logarithm is to any base, b, then

$$\frac{d}{dx}(\log_b x) = \frac{1}{x \cdot \ln b}$$

For example, if $g(x) = \log_5(x^2)$, then $g'(x) = \dfrac{1}{x^2 \cdot \ln 5} \cdot 2x = \dfrac{2x}{x^2 \ln 5}$ which reduces to $\dfrac{2}{x \ln 5}$.

Logarithmic Differentiation. There are still functions for which we have no differentiation formulas. For example, if both the base and the exponent are variables, we don't have any rules. A technique called logarithmic differentiation is often used. As the name implies, we use logarithms to help us. We can take the logarithm of both sides and then use the properties of logarithms.

A short review of some of the properties of logarithms:

$$\begin{aligned}
1)\ \log_b(xy) &= \log_b x + \log_b y \\
2)\ \log_b\left(\frac{x}{y}\right) &= \log_b x - \log_b y \\
3)\ \log_b x^p &= p \log_b x
\end{aligned}$$

Example 1: Derivatives of a Sum of Simple Exponential and Logarithmic Functions

If $f(x) = 3 \cdot 2^x + 4x^2 + 5 \ln x$, find $f'(x)$.

Strategy. We actually have 3 types of functions here held together with plus signs. First look at $3 \cdot 2^x$. Since the order of operations tells us that powers precede multiplication, it is not possible to multiply this out any further. Be sure you believe and understand that we do **NOT** have 6^x.

$$\frac{d}{dx}(3 \cdot 2^x) = 3 \cdot 2^x \cdot \ln 2$$

The 2nd term is a polynomial and uses the power rule.

$$\frac{d}{dx}(4x^2) = 8x$$

The 3rd term has a logarithm in it, and therefore we use the formula for the derivative of the logarithmic function.

$$\frac{d}{dx}(5 \ln x) = 5 \frac{d}{dx} \ln x = 5 \cdot \frac{1}{x} = \frac{5}{x}$$

Put this all together for the solution as follows:

Solution.

$$f'(x) = 3(\ln 2 \cdot 2^x) + 4(2x) + 5\frac{1}{x} = 3 \ln 2 \cdot 2^x + 8x + \frac{5}{x}$$

Summary. Actually the derivative of the exponential and logarithmic functions are the simplest to remember. When you see e, either as the base of an exponential or a logarithmic function, think *easy!*

Example 2: Chain Rule on an Exponential Function

Find:
$$\frac{d}{dx}\left(6e^{5x^2-5}\right)$$

Strategy. Here is a straightforward exponential function where the power also needs to be differentiated and hence, we use the chain rule. Remember to rewrite exactly what is given and then multiply by the derivative of the power.

Solution. The inside function is $u = 5x^2 - 5$, and the outside function is $y = 6e^u$. Using the chain rule, we get:

$$f'(x) = \left(6e^{5x^2-5}\right)10x = 60xe^{5x^2-5}$$

Summary. Once again, you see that the derivative of the exponential is straightforward.

2.7. DERIVATIVES OF EXPONENTIAL AND LOGARITHMIC FUNCTIONS

Example 3: Chain Rule on a Logarithmic Function

Compute the derivative of $f(x) = \ln\left(\sqrt{2x+1}\right)$.

Strategy. In this problem, we must remember that whenever we differentiate $\ln f(x)$, we start with $\dfrac{1}{f(x)}$ and then use the chain rule and differentiate $f(x)$ ending up with $\dfrac{f'(x)}{f(x)}$.

Solution. The inside function is $u = \sqrt{2x+1} = (2x+1)^{\frac{1}{2}}$, and the outside function is $y = \ln u$. Note that the derivative of the inside function also requires the chain rule. Applying the chain rule, we get:

$$\begin{aligned}
f'(x) &= \frac{1}{\sqrt{2x+1}} \cdot \frac{d}{dx}(2x+1)^{\frac{1}{2}} \\
&= \frac{1}{\sqrt{2x+1}} \cdot \frac{1}{2}(2x+1)^{\frac{-1}{2}} \cdot \frac{d}{dx}(2x+1) \\
&= \frac{1}{\sqrt{2x+1}} \cdot \frac{1}{2\sqrt{2x+1}} \cdot 2 \\
&= \frac{1}{2x+1}
\end{aligned}$$

Example 4: Logarithmic Differentiation

a) Find the derivative of $f(x) = x^{3x^2+4}$

b) Compute the derivative of $f(x) = (\cos x)^{2x+1}$, $-\frac{\pi}{2} < x < \frac{\pi}{2}$.

Strategy. If we see a variable in the base **and** the exponent of an expression, this is the method to choose. We will take the natural logarithm of each side of the equation and use the properties of logarithms to find $f'(x)$.

Solution. a) We begin by taking the natural logarithm of both sides of the equation $f(x) = x^{3x^2+4}$. Then we use the log property: $\log_b x^p = p \log_b x$.

$$\begin{aligned} \ln[f(x)] &= \ln(x^{3x^2+4}) \\ &= (3x^2+4)\ln x \end{aligned}$$

Next, differentiate both sides of this equation using the chain rule on the left and the product rule on the right.

$$\frac{1}{f(x)} \cdot f'(x) = (6x)(\ln x) + (3x^2+4)\left(\frac{1}{x}\right)$$

Solving this equation for $f'(x)$ and then replacing $f(x)$ with the beginning function, we get:

$$\begin{aligned} f'(x) &= \left[(6x \ln x) + (3x^2+4)\left(\frac{1}{x}\right)\right] \cdot f(x) \\ &= \left[(6x \ln x) + (3x^2+4)\left(\frac{1}{x}\right)\right] \cdot (x^{3x^2+4}) \end{aligned}$$

b) Once again, we begin by taking the natural logarithm of both sides of the equation $f(x) = (\cos x)^{2x+1}$. We have:

$$\begin{aligned} \ln[f(x)] &= \ln\left[(\cos x)^{2x+1}\right] \\ &= (2x+1)\ln(\cos x) \end{aligned}$$

Now differentiate both sides of this equation. On the left side we apply the chain rule, on the right side we use the product rule for the product of $(2x+1)$ and $\ln(\cos x)$ and the chain rule for $\ln(\cos x)$:

$$\begin{aligned} \frac{1}{f(x)} f'(x) &= 2\ln(\cos x) + (2x+1)\frac{1}{\cos x}(-\sin x) \\ &= 2\ln(\cos x) - (2x+1)\frac{\sin x}{\cos x} \end{aligned}$$

Solving this equation for $f'(x)$, we get:

$$\begin{aligned} f'(x) &= \left(2\ln(\cos x) - (2x+1)\frac{\sin x}{\cos x}\right) f(x) \\ &= \left(2\ln(\cos x) - (2x+1)\frac{\sin x}{\cos x}\right) (\cos x)^{2x+1} \end{aligned}$$

Summary. As you probably noticed, using logarithmic differentiation gives quite messy answers, but at least we can get an answer!

2.8. Implicit Differentiation and Inverse Trigonometric Functions

Key Topics.
- Differentiating Implicitly Defined Functions
- Derivatives of Inverse Trigonometric Functions

Worked Examples.
- Finding a Tangent Line by Implicit Differentiation
- Implicit Differentiation Involving a Product Rule
- Implicit Differentiation Involving a Chain Rule
- Derivative of an Inverse Trigonometric Function
- Derivative of an Inverse Trigonometric Function with a Quotient Rule

Overview

In this section we will study the derivatives of functions of x which cannot be explicitly stated. We will also study the derivatives of the inverse trigonometric functions.

Differentiating Implicitly Defined Functions. As stated in the text, many times we cannot explicitly solve an equation for y in terms of x. If the equation contains the letters x and y it is **implied** that y is a function of x.

Remember that when we differentiated any algebraic power function using the power rule and the chain rule, we brought the power in front, repeated the function, and then took the derivative of the inner function. Essentially, this tells us if $f(x) = x^3$, then $f'(x) = 3x^2 \cdot \frac{dx}{dx}$. Since $\frac{dx}{dx} = 1$, (just like any other quotient where the numerator and the denominator are the same and not zero) and multiplication by 1 doesn't change anything, we usually do not write the $\frac{dx}{dx}$.

However, using the same reasoning, if $f(x) = y^3$, then $f'(x) = 3y^2 \frac{dy}{dx}$. Since it cannot be simplified, the $\frac{dy}{dx}$ is absolutely necessary. It is truly just another way to look at the chain rule.

Derivatives of Inverse Trigonometric Functions. Many applications involve the inverse trig functions. Let's look at one way to derive these derivatives.

If $y = \sin^{-1} x$, then by definition of the inverse sine function this means that $\sin y = x$. This looks a little strange, because it tells us that y is an **angle**. Sketch a right triangle with angle y in the lower left corner. Since the sine function is $\frac{\text{opposite side}}{\text{hypotenuse}}$, we can write $\sin y = \frac{x}{1}$, and we can label all three sides of the triangle, using the Pythagorean Theorem for the third side.

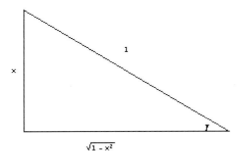

Find $\dfrac{dy}{dx}$ if $y = \sin^{-1} x$. Rewrite as $\sin y = x$. Differentiating both sides of the equation using implicit differentiation, we get:

$$\frac{d}{dx} \sin y = \frac{d}{dx}(x)$$
$$(\cos y)\frac{dy}{dx} = 1$$
$$\frac{dy}{dx} = \frac{1}{\cos y}$$

Then using the triangle and the definition of $\cos y = \dfrac{\text{adjacent side}}{\text{hypotenuse}} = \dfrac{\sqrt{1-x^2}}{1}$, we can substitute and arrive at the derivative:

$$\frac{dy}{dx} = \frac{1}{\sqrt{1-x^2}}$$

In a similar manner we can derive the derivatives of the other inverse trig functions. Happily, the derivatives of $y = \cos^{-1} x$, $y = \cot^{-1} x$, and $y = \csc^{-1} x$ are just the negatives of the other three inverse trigonometric functions.

Example 1: Finding a Tangent Line by Implicit Differentiation

Find an equation of the tangent line for the ellipse $x^2 + 2y^2 = 9$ at the point $(1, 2)$.

Strategy. Just like we can add a term to both sides of an equation, we can differentiate both sides of an equation. Remember, also, that the derivative of a constant is 0.

Using the chain rule:
$$\frac{d}{dx}(x^2) = 2x\frac{dx}{dx} = 2x$$
From now on we will not write $\frac{dx}{dx}$ every time.

$$\frac{d}{dx}(2y^2) = 4y\frac{dy}{dx}$$
$$\frac{d}{dx}(9) = 0$$

Put all this together to get the solution.

Solution. Differentiate both sides of the equation $x^2 + 2y^2 = 9$ with respect to x.

$$\frac{d}{dx}\left(x^2 + 2y^2\right) = \frac{d}{dx}(9)$$
$$2x + 2(2y)\frac{dy}{dx} = 0$$
$$2x + 4y\frac{dy}{dx} = 0$$

Solve the last equation for $\frac{dy}{dx}$, and substitute $x = 1$, and $y = 2$, in order to find the slope of the tangent line at the point $(1, 2)$.

$$4y\frac{dy}{dx} = -2x$$
$$\frac{dy}{dx} = \frac{-2x}{4y} = \frac{-x}{2y}$$

Then evaluating $\frac{dy}{dx}$ at the point $(1, 2)$ we have $\frac{dy}{dx} = \frac{-1}{4}$.

The equation of the tangent line with slope $-\frac{1}{4}$ through the point $(1, 2)$ is given by:

$$y - 2 = -\frac{1}{4}(x - 1)$$

$$y = -\frac{1}{4}(x-1) + 2$$
$$y = -\frac{1}{4}x + \frac{1}{4} + 2$$
$$y = -\frac{1}{4}x + \frac{9}{4}$$

Here is a picture of this implicitly defined function with the tangent line at the point $(1, 2)$.

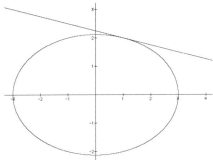

Summary. If it helps you to write $\dfrac{dx}{dx}$ so that the pattern for all terms is identical, then keep doing that for as long as you need. It is completely correct.

Example 2: Implicit Differentiation Involving a Product Rule

Find $\frac{dy}{dx}$ for $x^2 - xy + y^2 = 7$. Then find the slope of the tangent line at the point $(3, 1)$.

Strategy. We will approach this problem by differentiating both sides just as we did in example 1. The middle term, however, $(-xy)$ makes this a more complicated problem. It is a product and therefore we need to use the product rule. Also, we need to take extreme caution with the negative sign.

$$\frac{d}{dx}(x^2) = 2x$$
$$\frac{d}{dx}(-xy) = -[y\frac{dx}{dx} + x\frac{dy}{dx}] = -[y + x\frac{dy}{dx}] = -y - x\frac{dy}{dx}$$
$$\frac{d}{dx}(y^2) = 2y\frac{dy}{dx}$$
$$\frac{d}{dx}(7) = 0$$

Put this all together to get the solution.

Solution.

$$\frac{d}{dx}(x^2 - xy + y^2) = \frac{d}{dx}(7)$$
$$2x - y - x\frac{dy}{dx} + 2y\frac{dy}{dx} = 0$$

Next, keep the terms with $\frac{dy}{dx}$ on the left side and bring the other terms to the right side.

$$2y\frac{dy}{dx} - x\frac{dy}{dx} = y - 2x$$

Factor out the common factor of $\frac{dy}{dx}$ and then divide for the final answer.

$$\frac{dy}{dx}(2y - x) = y - 2x$$
$$\frac{dy}{dx} = \frac{y - 2x}{2y - x}$$

Substitute $x = 3$, $y = 1$ to find the slope at the point $(3, 1)$.

$$\frac{dy}{dx} = \frac{1 - 2(3)}{2(1) - 3} = \frac{-5}{-1} = 5$$

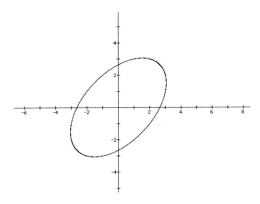

Summary. Whenever you are asked for slope, you need to find the derivative. Also, be very careful whenever you have a product with a negative sign in front of it.

Example 3: Implicit Differentiation Involving a Chain Rule

Find $\dfrac{dy}{dx}$ for $\sin(xy) = \sin x + \sin y$.

Strategy. Once again, we begin by differentiating both sides taking care to use the product rule for the xy on the left hand side.

Solution. Differentiate both sides of the equation $\sin(xy) = \sin x + \sin y$ with respect to x. On the left side use the product rule to differentiate the product $x \cdot y$, and then the chain rule to differentiate $\sin(xy)$.

$$\frac{d}{dx}(\sin(xy)) = \frac{d}{dx}(\sin x + \sin y)$$

$$\cos(xy)\left(y + x\frac{dy}{dx}\right) = \cos x + (\cos y)\frac{dy}{dx}$$

Now we solve this equation for $\dfrac{dy}{dx}$.

$$y\cos(xy) + x\cos(xy)\frac{dy}{dx} = \cos x + \cos y \frac{dy}{dx}$$

$$x\cos(xy)\frac{dy}{dx} - \cos y\frac{dy}{dx} = \cos x - y\cos(xy)$$

$$(x\cos(xy) - \cos y)\frac{dy}{dx} = \cos x - y\cos(xy)$$

$$\frac{dy}{dx} = \frac{\cos x - y\cos(xy)}{x\cos(xy) - \cos y}$$

Below is a picture of the implicitly defined function $\sin(xy) = \sin x + \sin y$.

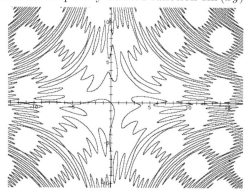

Summary. As you can see, you must work slowly and carefully so that you do not leave out any of the parts. You have trig. functions, product rule and chain rule here along with some factoring from algebra. A lot happens in this interesting problem.

2.8. IMPLICIT DIFFERENTIATION AND INVERSE TRIGONOMETRIC FUNCTIONS

Example 4: Derivative of an Inverse Trigonometric Function

Compute the derivative of $f(x) = \tan^{-1}\left(\sqrt{2x+1}\right)$.

Strategy. When we look at the formula for the derivative of the inverse tangent function we see:
$$\frac{d}{du}\tan^{-1} u = \frac{1}{1+u^2}$$

If $u = f(x)$, then $1 + u^2 = 1 + (f(x))^2$. For example if we have
$$f(x) = \tan^{-1}\left(5x^2\right)$$
then
$$f'(x) = \frac{1}{1+(5x^2)^2} \cdot 10x = \frac{10x}{1+25x^4}$$
Note the parentheses around the $5x^2$.

In the given problem, we have:
$$f(x) = \tan^{-1}\left(\sqrt{2x+1}\right) = \tan^{-1}\left((2x+1)^{\frac{1}{2}}\right)$$

The rest of the solution follows.

Solution. From the chain rule, we get:
$$\begin{aligned}
f'(x) &= \frac{1}{1+\left(\sqrt{2x+1}\right)^2} \cdot \frac{d}{dx}\left((2x+1)^{\frac{1}{2}}\right) \\
&= \frac{1}{1+(2x+1)} \cdot \frac{1}{2}(2x+1)^{\frac{-1}{2}} \cdot 2 \\
&= \frac{1}{(2+2x)} \frac{1}{\sqrt{2x+1}} \\
&= \frac{1}{(2+2x)(\sqrt{2x+1})}
\end{aligned}$$

Summary. It is not possible to overstress the importance of using parentheses. Whenever you substitute more than one symbol for one symbol (i.e.: -2 for x) you MUST enclose the substitution in parentheses!

Example 5: Derivative of an Inverse Trigonometric Function with a Quotient Rule

---

Compute the derivative of
$$f(x) = \frac{\sin^{-1} x}{\sqrt{1-x^2}}$$

Strategy. There are many steps in this problem. We will work slowly showing one step at a time. We must use the quotient rule, the chain rule and the derivative of the inverse sine function.

Solution. From the quotient rule, we have

$$f'(x) = \frac{\sqrt{1-x^2}\frac{d}{dx}\left(\sin^{-1} x\right) - \sin^{-1} x \frac{d}{dx}\left(\sqrt{1-x^2}\right)}{\left(\sqrt{1-x^2}\right)^2}$$

$$= \frac{\sqrt{1-x^2} \cdot \frac{1}{\sqrt{1-x^2}} - (\sin^{-1} x)\left(\frac{1}{2}(1-x^2)^{\frac{-1}{2}} \cdot 2x\right)}{1-x^2}$$

$$= \frac{\frac{\sqrt{1-x^2}}{\sqrt{1-x^2}} - \frac{-\sin^{-1} x}{\sqrt{1-x^2}}}{1-x^2}$$

To simplify the complex fraction, continue.

$$= \frac{\frac{\sqrt{1-x^2} + \sin^{-1} x}{(1-x^2)^{\frac{1}{2}}}}{(1-x^2)}$$

$$= \frac{\sqrt{1-x^2} + \sin^{-1} x}{(1-x^2)^{\frac{1}{2}}} \cdot \frac{1}{(1-x^2)^1}$$

$$= \frac{\sqrt{1-x^2} + \sin^{-1} x}{(1-x^2)^{\frac{3}{2}}}$$

Summary. As you can see in this problem, many times the algebraic simplification is a lot trickier than the calculus!

2.9. The Mean Value Theorem

Key Topics.
- The Mean Value Theorem
- Using the Sign of the First Derivative to Determine if a Functions is Increasing or Decreasing
- Antiderivatives - Finding Every Function with a Given Derivative

Worked Examples.
- Determining the Number of Zeros of a Function
- A Mean Value Theorem Example
- Determine when a Function is Increasing and when it is Decreasing
- Finding Every Function with a Given Derivative

Overview

In section 2.9 we will study three important topics in calculus. The first is called the Mean Value Theorem. The word "mean" in math is usually used to convey an average. It is not exactly an average here, but it does talk about a number somewhere in between two other numbers on an interval.

Secondly, we will use the fact that a derivative gives us a slope and use the derivative to learn more about the behavior of a function.

And, lastly, in this section, we will have a little taste of the other branch of calculus - antidifferentiation, or integration.

The Mean Value Theorem. If you drive a car for two hours and cover 110 miles, then you know that the average speed is 55 m.p.h. The Mean Value Theorem (MVT) tells us that at some point you were actually going 55 m.p.h.

Mathematically, this theorem relates the average rate of change of a function on an interval to the instantaneous rate of change at a specific point in that interval. Geometrically, the MVT is quite easy to see. Take any graph which satisfies the conditions of the MVT, which are:

$f(x)$ is continuous on a closed interval $[a, b]$ and differentiable on the open interval (a, b)

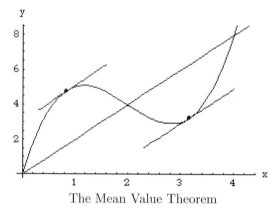
The Mean Value Theorem

Now find these points on the graph: $(a, f(a))$ and $(b, f(b))$. Draw the secant line between these two points. The slope of this secant line is:

$$m = \frac{y_2 - y_1}{x_2 - x_1} = \frac{f(b) - f(a)}{b - a}$$

What the MVT tells us is that if we draw tangent lines to our curve at every point between $x = a$ and $x = b$, then at least one of those tangent lines will be parallel to the secant line we drew. Call this point c. We know that slope is equal to the derivative. Hence, $f'(c)$ gives us the slope at the point where $x = c$, which is parallel to the secant line.

$$f'(c) = \frac{f(b) - f(a)}{b - a}$$

A direct result of the MVT is **Rolle's Theorem**, which puts one more major condition on the MVT. It says that $f(a) = f(b)$ which, of course, makes the secant line horizontal and thus has a slope of zero. Rolle's Theorem says, then, that at some point c between a and b (i.e.: $a < c < b$), the tangent line is horizontal.

Using the Sign of the First Derivative to Determine if a Function is Increasing or Decreasing. Since the slope is equal to the derivative, if we have a positive slope at every point x on an interval, $(f'(x) > 0)$, then the function is increasing on that interval. Similarly, if we have a negative slope at every point on an interval, ($f'(x) < 0$), then the function is decreasing on that interval.

Antiderivatives - Finding Every Function with a Given Derivative. Just like subtraction *undoes* addition, and division *undoes* multiplication, you probably have figured out that there is a process which *undoes* differentiation. That process is appropriately called anti-differentiation.

Example 1: Determining the Number of Zeros of a Function

Show that the equation $x^3 + 2x + 2 = 0$ has exactly one solution.

Strategy. The *easy* part here is to show that there is a solution. The *harder* part is to show that there is only one solution. We are going to use Rolle's Theorem to accomplish that.

We will assume that there is more than one zero and show that this assumption leads to a contradiction, that is it leads to a conclusion which cannot be true.

Solution. Set $f(x) = x^3 + 2x + 2$. f is a continuous, and differentiable function, and thus satisfies the hypothesis of Rolle's Theorem.

We first make sure that $f(x) = 0$ does have a solution. We will find an x where $f(x) > 0$ and another x where $f(x) < 0$. Then, since the function is continuous, we know that it must cross the x-axis at least once between the two points.

$$
\begin{aligned}
f(0) &= 2, \text{ a positive number} \\
f(-1) &= -1, \text{ a negative number}
\end{aligned}
$$

Since f is continuous f has to be equal to 0 for some x between -1 and 0.

Why can't $f = 0$ have more than one solution?

Suppose f has more than one solution. That means we can find a and b with $f(a) = f(b) = 0$. Since f is continuous and differentiable everywhere, in particular on the interval $[a, b]$, we get from Rolle's Theorem, that there is a number c with $f'(c) = 0$. But $f'(x) = 3x^2 + 2$ and there is no number which can make this equal to zero. Therefore our assumption of two numbers a and b must have been incorrect. We can conclude that f has precisely one zero.

Note that we could have also used Theorem 9.2 to do this problem.

Summary. Proof by contradiction is a common technique. If you make an assumption and show that it leads to a conclusion which contradicts what you know to be true, then the assumption must be false.

Also, note that the MVT and Rolle's Theorem tell us the existence of a point. They do not tell us how to find that point.

Example 2: A Mean Value Theorem Example

Show that the function $f(x) = x^2 + 2x - 1$ satisfies the hypotheses of the Mean Value Theorem on the interval $[0, 1]$, and find the value of c in $(0, 1)$ satisfying:

$$f'(c) = \frac{f(b) - f(a)}{b - a}$$

Strategy. We can find the slope of the secant line between $(0, f(0))$ and $(1, f(1))$ relatively easily. Then all we have to do is find the value, c, such that $(0 < c < 1)$ where $f'(c)$ is equal to that slope.

Solution. $f(x) = x^2 + 2x - 1$ is a continuous, and differentiable function, thus satisfying the hypothesis of the Mean Value Theorem. By the Mean Value Theorem there is a c in $(0, 1)$ with

$$f'(c) = \frac{f(b) - f(a)}{b - a}$$

Since $f(b) = f(1) = 2$, $f(a) = f(0) = -1$, and $f'(x) = 2x + 2$, the MVT becomes:

$$\begin{aligned} f'(c) &= \frac{f(1) - f(0)}{1 - 0} \\ 2c + 2 &= \frac{2 - (-1)}{1} \\ 2c + 2 &= 3 \\ 2c &= 1 \\ c &= \frac{1}{2} \end{aligned}$$

Summary. We have found the value $c = \frac{1}{2}$, which lies in the interval $(0, 1)$, and where the tangent line is parallel to the secant line through the points $(0, -1)$ and $(1, 2)$.

Example 3: Determine when a Function is Increasing and when it is Decreasing

Determine the intervals on which $f(x) = 2x^3 - 3x^2 - 12x$ is increasing, and on which intervals f is decreasing.

Strategy. Here we will use the derivative, which is the slope, to determine the intervals on which this function is increasing and on which intervals it is decreasing.

Solution. First we need to find the x-values which make the derivative equal to zero. (i.e.: solve $f'(x) = 0$)

$$f'(x) = 6x^2 - 6x - 12 = 6(x^2 - x - 2) = 6(x+1)(x-2) = 0$$

$$x = -1 \text{ and } x = 2$$

Draw a number line and put marks at $x = -1$ and at $x = 2$:

We need to see if the derivative is positive or negative on each of these three intervals which we formed.

If $x < -1$, then $6(x+1)(x-2) < 0$ and the function is **decreasing.**
If $-1 < x < 2$, then $6(x+1)(x-2) > 0$ and the function is **increasing.**
If $x > 2$, then $6(x+1)(x-2) < 0$ and the function is **decreasing.**

Put these findings on the number line.

Summary. When we are looking at a specific interval, it helps to choose *any* number in that interval to test the derivative. For example, when we check $x < -1$, we can choose $x = -2$ and find that $f''(-2) = 6(-2+1)(-2-2) > 0$. It doesn't matter which number we choose on the interval. Think about why it doesn't matter.

Example 4: Finding Every Function with a Given Derivative

Find all functions $g(x)$ with $g'(x) = 10x^5 - \frac{2}{x^2} + \frac{1}{\sqrt{x}}$.

Strategy. Do not let problems like this scare you. You will have formulas to work these later in the course. Here we will use logic. We need to *undo* what was done in order to get the derivative.

Solution. If $g'(x) = 10x^5$, we know that the power of 5 came from going down one power. It, therefore, started out with a power of 6. So we know we had x^6 in the original function. The 6 was brought in front and multiplied. To undo that, we need to divide. So we can see that

an antiderivative of $10x^5$ is $10\left(\frac{1}{6}x^6\right) = \frac{5}{3}x^6$.

Use similar logic for $-2x^{-2}$ and $x^{\frac{-1}{2}}$:

An antiderivative of $-\frac{2}{x^2} = -2x^{-2}$ is $-2\left(\frac{1}{-1}x^{-1}\right) = 2x^{-1} = \frac{2}{x}$.

An antiderivative of $\frac{1}{\sqrt{x}} = x^{-\frac{1}{2}}$ is $\frac{2}{1}x^{\frac{1}{2}} = 2\sqrt{x}$

All functions $g(x)$ with $g'(x) = 10x^5 - \frac{2}{x^2} + \frac{1}{\sqrt{x}}$ are given by $g(x) = \frac{5}{3}x^6 + \frac{2}{x} + 2\sqrt{x} + c$.

Summary. Since the problem asks for **every** function with the given derivative we **MUST** add $+c$ to our answer. The c stands for any constant. Since $\frac{d}{dx}(c) = 0$ for all c, it doesn't change the derivative. The $+c$ is necessary whenever we find an antiderivative.

CHAPTER 3

Applications of Differentiation

3.1. Linear Approximation and Newton's Method

Key Topics.
- Linear Approximation
- Differentials
- Newton's Method

Worked Examples.
- Using Linear Approximation to Approximate Square Roots
- Using a Linear Approximation to Perform Linear Interpolation
- Using Newton's Method to Approximate a Square Root
- An Example when Newton's Method Fails to Work

Overview

One of the reasons we use a calculator is to compute values of transcendental functions. We can find $\sin \frac{\pi}{4}$, for example, by using a unit circle and/or a right triangle. We can do so for the trig functions of *special angles*. However, most values for the angle do not lend themselves to the simple drawing of a triangle. Did you ever wonder where the answer came from when you pressed the button on a calculator and got a long decimal? Or, how would we go about getting those answers? Somebody has to program the calculators and the computers. What follows are several ways in which we can closely approximate what we cannot exactly compute.

Linear Approximation. Linear Approximation is actually another name for tangent line approximation. When you draw a tangent line to a curve, that line *hugs* the curve **in the vicinity** of the point of tangency. If we have the equation of the line at the point of tangency, we can use the equation to approximate the value of the function which is close to that point.

If you understand the way that the equation for linear approximation is derived, it will help you understand it better.

Given a graph, $y = f(x)$, the slope of the tangent line at x_0 is given by $f'(x_0)$, and the point $(x_0, f(x_0))$ is a point on this line. We use the point-slope formula.

$$\begin{aligned} y - f(x_0) &= f'(x_0)(x - x_0) \\ y &= f(x_0) + f'(x_0)(x - x_0) \end{aligned}$$

Since this is the equation of the tangent line to the graph, we will call it $L(x)$ and write:

$$L(x) = f(x_0) + f'(x_0)(x - x_0)$$

If you remember that the derivative is equal to the slope, then $f'(x_0) = m$ and the equation doesn't look so different anymore!

Differentials. Differentials of x and y measure the difference between one value of x or y and another. Change in y is written Δy (read delta y or change in y) and represents an incremental change. This change is very small. If

$$\begin{aligned} \Delta x &= x_1 - x_0 = \text{change in } x \\ \Delta y &= y_1 - y_0 = f(x_1) - f(x_0) = \text{change in } y \end{aligned}$$

We can think of $dx = \Delta x$, and $dy =$ approximate change in y, then

$$dy = f'(x_0) dx$$

Newton's Method. We can use Newton's method as a way to approximate the zeros of a function, that is the x-values for which $y = f(x) = 0$. We start by making a guess as to what the zero is. This must be an *educated guess* - it must be close to the actual answer. For example, if you can see from the graph that the answer lies between 3 and 4, choose $x = 3$ or $x = 4$ as the starting place. We call the starting number x_0. Next, we find the equation of the tangent line at x_0. Using the equation developed under Linear Approximation, we have:

$$y = f(x_0) + f'(x_0)(x - x_0)$$

We need the x-intercept of this equation, which should be closer to the answer than our original guess. We call the x-intercept, x_1. As always, we find the x-intercept by setting $y = 0$ and solving for x_1.

$$\begin{aligned} 0 &= f(x_0) + f'(x_0)(x_1 - x_0) \\ -f(x_0) &= f'(x_0)(x_1 - x_0) \\ \frac{-f(x_0)}{f'(x_0)} &= x_1 - x_0 \\ x_1 &= x_0 - \frac{f(x_0)}{f'(x_0)} \end{aligned}$$

Then we start all over again with x_1 being our guess and x_2 being the next x-intercept which should be even closer to the correct answer.

$$x_2 = x_1 - \frac{f(x_1)}{f'(x_1)}$$

Each time we repeat this, we get closer to the zero we are looking for. Each step is called an *iteration*.

Example 1: Using Linear Approximation to Approximate Square Roots

Use linear approximation to estimate $\sqrt{4.2}$ and $\sqrt{3.996}$.

Strategy. We really only need to get one equation here because both $\sqrt{4.2}$ and $\sqrt{3.996}$ are very close to $\sqrt{4}$ which we know.

Note: What we do for any linear approximation is to find a point $x = a$ close to the ones we want to approximate. For this problem, we use $a = 4$ to approximate values close to $\sqrt{4}$.

Solution. We first need to choose a function. Since we are approximating square roots, we choose $f(x) = \sqrt{x}$. We are approximating square roots close to the square root of 4, therefore we can choose $x_0 = 4$. First, find $f(4)$, and $f'(4)$.

$$f(x) = \sqrt{x} = x^{\frac{1}{2}} \implies f'(x) = \frac{1}{2}x^{\frac{-1}{2}} = \frac{1}{2\sqrt{x}}$$

$$f(4) = 2 \implies f'(4) = \frac{1}{4}$$

Then use these values to find the linear approximation $L(x)$.

$$\begin{aligned} L(x) &= f(x_0) + f'(x_0)(x - x_0) \\ &= f(4) + f'(4)(x - 4) \\ &= 2 + \frac{1}{4}(x - 4) \end{aligned}$$

You might be tempted to simplify this equation of the tangent line further. However, simplifying would make our later computations more difficult. Why? Because when we plug in values for x later on, the expression $\frac{1}{4}(x - 4)$ will be easier to evaluate than $\frac{1}{4}x - 1$.

$$f(x) \approx L(x) = 2 + \frac{1}{4}(x - 4) \text{ for } x \text{ close to } x_0 = 4$$

We can now use this to approximate the two square roots. For the first square root we substitute $x = 4.2$, and for the second $x = 3.996$ into the linear approximation $L(x)$.

$$\sqrt{4.2} \approx 2 + \frac{1}{4}(4.2 - 4) = 2 + \frac{1}{4}(0.2) = 2 + 0.05 = 2.05$$

$$\sqrt{3.996} \approx 2 + \frac{1}{4}(3.996 - 4) = 2 + \frac{1}{4}(-0.004) = 2 - 0.001 = 1.999$$

Summary. If you work with this for a while, you will find that you are actually working with the equation of a line. The different notation makes it a little harder, so make sure you take the time needed to totally understand this.

Example 2: Using a Linear Approximation to Perform Linear Interpolation

Suppose $C(T)$ is the cost to heat Alison's dorm room in cents per day, where T is the outside temperature.

T (outside temperature in °F)	20	25	30	35
$C(T)$ (heating cost in cents per day)	85	75	66	60

Estimate the heating cost when the outside temperature is 22 °F.

Strategy. Once again, we start with a T-value for which we know the function value - this time from the table of given values. Since we are interested in $T = 22\,°\text{F}$, we start with the closest T-value which is $T = 20\,°\text{F}$ as follows.

Solution. The T-value in the table closest to $T = 22\,°\text{F}$ is given by $T = 20\,°\text{F}$. We know from the table that $C(20) = 85$. The linear approximation of $C(T)$ at $T_0 = 20$ is given by

$$\begin{aligned} L(T) &= C(T_0) + C'(T_0)(T - T_0) \\ &= C(20) + C'(20)(T - 20) \end{aligned}$$

We need to estimate $C'(20)$. There are many different ways to approximate $C'(20)$, but the straight forward approximation is to find the average rate of change of C between $T = 20$ and $T = 25$.

$$\begin{aligned} C'(20) &\approx \frac{C(25) - C(20)}{25 - 20} \frac{\text{cents per day}}{°\text{F}} \\ &= \frac{75 - 85}{5} = -\frac{10}{5} = -2 \text{ cents per day per } °\text{F} \end{aligned}$$

$$L(T) = 85 - 2(T - 20)$$

Thus, for T close to $20\,°\text{F}$, $C(T) \approx L(T) = 85 - 2(T - 20)$, and

$$L(22) = 85 - 2(22 - 20) = 85 - 2 \cdot 2 = 81 \text{ cents per day}$$

The approximate cost to heat Alison's dorm room when the outside temperature is $22\,°\text{F}$ is approximately 81 cents per day.

How does this example relate to **differentials**? In this example:

$$\Delta T = 2, C'(20) \approx -2$$
$$\Delta C \approx dC = C'(20) \cdot \Delta T = -2\Delta T = -2(2) = -4 \text{ cents per day}$$

The approximate change in cost is -4 cents per day, therefore the cost is approximately the cost at $T = 20$ which is 85 cents, plus the change in cost of -4 cents, for an estimated cost of 81 cents a day.

Summary. Once again we started with a value we knew and were able to approximate the value we wanted. If you are confused with the units, let's rewrite the equation with units:

$$C'(20) = \frac{-2 \text{ cents per day}}{{}^\circ F}$$

Then in the equation, this is multiplied by $(x-2)$ which is in ${}^\circ F$. Consequently, we have:

$$\left(\frac{-2 \text{ cents per day}}{{}^\circ F}\right) \cdot \frac{(x-2)\,{}^\circ F}{1}$$

The ${}^\circ F$ cancel, leaving cents per day!

Example 3: Using Newton's Method to Approximate a Square Root

Use Newton's Method to approximate $\sqrt{2}$.

Strategy. Newton's Method approximates zeros of functions. In other words, it finds the x-values when $y = 0$, which are also called the x-*intercepts* or the *roots*. So first we need an function where $\sqrt{2}$ is a zero. $f(x) = x^2 - 2$ is the function which comes to mind.

$$\begin{aligned} x^2 - 2 &= 0 \\ x^2 &= 2 \\ x &= \pm\sqrt{2} \end{aligned}$$

Solution. Since $f(x) = x^2 - 2$, we will have $f'(x) = 2x$ in our calculations. A good initial approximation of $\sqrt{2}$ is $x_0 = 1$. $f(1) = -1$, and $f'(1) = 2$. Using Newton's method we have:

$$\begin{aligned} x_1 &= x_0 - \frac{f(x_0)}{f'(x_0)} \\ &= 1 - \frac{-1}{2} = 1 + \frac{1}{2} = \frac{3}{2} = 1.5 \end{aligned}$$

Continuing this process, we have $f\left(\frac{3}{2}\right) = \left(\frac{3}{2}\right)^2 - 2 = \frac{9}{4} - 2 = \frac{1}{4}, f'\left(\frac{3}{2}\right) = 2 \cdot \frac{3}{2} = 3$

$$\begin{aligned} x_2 &= x_1 - \frac{f(x_1)}{f'(x_1)} \\ &= \frac{3}{2} - \frac{\frac{1}{4}}{3} \qquad\qquad (\frac{\frac{1}{4}}{3} = \frac{\frac{1}{4}}{\frac{3}{1}} = \frac{1}{4} \cdot \frac{1}{3}) \\ &= \frac{3}{2} - \frac{1}{4} \cdot \frac{1}{3} \\ &= \frac{3}{2} - \frac{1}{12} \qquad\qquad \text{(common denominator)} \\ &= \frac{18 - 1}{12} = \frac{17}{12} \approx 1.4167 \end{aligned}$$

Summary. Using a calculator, we find $\sqrt{2} \approx 1.4142$. We can continue with Newton's method if we want to get even closer, but as you can see, it is quite tedious.

Example 4: An Example when Newton's Method Fails to Work

Explain why Newton's Method fails to work for $f(x) = x^2 - 2$, and $x_0 = 0$.

Strategy. As you will see from this problem Newton's method doesn't always work. If it is going to work, the tangent line must have an x-intercept and you must choose a starting point close to the x-intercept.

Solution. Starting with $f(x) = x^2 - 2$, we have the values $f'(x) = 2x$, $f(0) = -2$, and $f'(0) = 0$. Newton's method gives

$$\begin{aligned} x_1 &= x_0 - \frac{f(x_0)}{f'(x_0)} \\ &= 0 - \frac{-2}{0} \text{ which is undefined} \end{aligned}$$

Newton's method fails to work since x_1 is undefined. But why is it undefined?

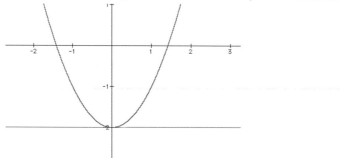

A horizontal tangent line

Newton's method finds the x-intercept of the tangent line at a given point. However, at $x = 0$, the tangent line to the parabola at $x = 0$ is parallel to the x-axis and therefore has no x-intercepts.

Summary. Newton's method does not always work. The reason why Newton's method does not work could be a horizontal tangent line as seen in this example, it could be that we chose our starting point too far away from a zero, or it could approach a zero different from the chosen one. There are many different reasons why Newton's method might fail to work. When it works, it approaches the zero of the function very quickly.

3.2. Indeterminate Forms and L'Hôpital's Rule

Key Topics.
- L'Hôpital's Rule for Limits of Type $\frac{0}{0}$ and $\frac{\infty}{\infty}$
- L'Hôpital's Rule for Limits of Type $0 \cdot \infty$
- L'Hôpital's Rule for Limits of Type $\infty - \infty$
- L'Hôpital's Rule for Limits of Type $0^0, \infty^0, 1^\infty$

Worked Examples.
- A Limit of Type $\frac{0}{0}$
- A Limit of Type $\frac{\infty}{\infty}$
- A Limit Requiring the Use of L'Hôpital's Rule Twice
- A Limit of Type $0 \cdot \infty$
- A Limit of Type $\infty - \infty$
- A Limit of Type 1^∞
- A Limit of Type 0^0

Overview

In chapter 1 there were many limits where we couldn't simply plug in the number to find the limit. If we did try, many times we got $\frac{0}{0}$ which is undefined. For example:

$$\lim_{x \to 1} \frac{x^2 - 1}{x - 1}$$

If we try to substitute 1 for x, we end up with $\frac{0}{0}$. In this case you factored and finished the problem as:

$$\lim_{x \to 1} \frac{(x-1)(x+1)}{(x-1)}$$
$$= \lim_{x \to 1} (x+1) = 2$$

Unfortunately, we cannot always factor, especially when there are transcendental functions.

L'Hôpital's Rule for Limits of Type $\frac{0}{0}$ and $\frac{\infty}{\infty}$. L'Hôpital published the very first calculus book and included the rule which follows. Even though he didn't come up with the rule, it is named after him since he was the one to publish it in a book! It says that if a limit is of the indeterminate form, $\frac{0}{0}$ or $\frac{\infty}{\infty}$, then we can take the derivative of the top and the derivative of the bottom, and then try substituting again.

L'Hôpital's Rule for Limits of Type $0 \cdot \infty$. In this case, we need to use algebraic properties to rewrite the problem as a quotient and then we can use L'Hôpital's Rule.

L'Hôpital's Rule for Limits of Type $\infty - \infty$. As in $0 \cdot \infty$, we will rewrite the expression as a quotient. This usually involves combining two fractions.

L'Hôpital's Rule for Limits of Type $0^0, \infty^0, 1^\infty$. This is the trickiest type. Because there is a variable in the exponent, we cannot work with it until we get the variable out of the exponent. The way we **always** accomplish this is by taking the natural log (ln) on both sides. Then we can use the property of logarithms that tells us that "exponents jump over logs".

$$\ln a^x = x \ln a$$

Example 1: A Limit of Type $\frac{0}{0}$

Find the limit.
$$\lim_{x \to 1} \frac{\ln x}{x - 1}$$

Strategy. First we must check and make sure that the criteria for L'Hôpital's Rule is fulfilled. If we substitute 1 for x, we end up with $\frac{0}{0}$, since $\ln 1 = 0$, and $1 - 1 = 0$. Use L'Hôpital's Rule which says to take the derivative of the top and the derivative of the bottom and then substitute 1 for x.

Solution. This limit has indeterminate form $\frac{0}{0}$ ($\ln 1 = 0$). L'Hôpital's rule gives

$$\lim_{x \to 1} \frac{\ln x}{x - 1} = \lim_{x \to 1} \frac{\frac{1}{x}}{1} = \frac{\frac{1}{1}}{1} = 1$$

Summary. See - L'Hôpital's Rule is really nice! The two derivatives here are straightforward, and we only had to use L'Hôpital's Rule once.

Example 2: A Limit of Type $\frac{\infty}{\infty}$

Find the limit.
$$\lim_{x \to \infty} \frac{\ln x}{x^2}$$

Strategy. This problem gives $\frac{\infty}{\infty}$ if we try to substitute. Clearly, we need to use L'Hôpital's Rule.

Solution. This limit has indeterminate form $\frac{\infty}{\infty}$. L'Hôpital's rule gives

$$\begin{aligned}
\lim_{x \to \infty} \frac{\ln x}{x^2} &= \lim_{x \to \infty} \frac{\frac{1}{x}}{2x} = \lim_{x \to \infty} \frac{\frac{1}{x}}{\frac{2x}{1}} \quad \text{(invert denominator)} \\
&= \lim_{x \to \infty} \left(\frac{1}{x}\right)\left(\frac{1}{2x}\right) \\
&= \lim_{x \to \infty} \frac{1}{2x^2} = 0
\end{aligned}$$

Summary. Remember if the denominator gets very large and the numerator remains constant, then the limit is zero.

Example 3: A Limit Requiring the Use of L'Hôpital's Rule Twice

Find the limit.
$$\lim_{t \to 0} \frac{\cos t - 1}{e^t - t - 1}$$

Strategy. Since $\cos 0 = 1$ and $e^0 = 1$, both numerator and denominator approach zero as t approaches zero. We need to use L'Hôpital's Rule twice in this problem.

Solution. This limit has indeterminate form $\frac{0}{0}$. Applying L'Hôpital's rule once gives

$$\lim_{t \to 0} \frac{\cos t - 1}{e^t - t - 1} = \lim_{t \to 0} \frac{-\sin t}{e^t - 1}$$

Since $\sin 0 = 0$ and $e^0 = 1$, we have the indeterminate form $\frac{0}{0}$ again. In this case applying L'Hôpital's rule once more gives

$$= \lim_{t \to 0} \frac{-\cos t}{e^t} = \frac{-\cos 0}{e^0} = \frac{-1}{1} = -1$$

Summary. You notice that we checked for the criteria for L'Hôpital's Rule each time it was used. A common mistake is to keep taking the derivative of the top and the derivative of the bottom without checking the criteria **every** time you use L'Hôpital's Rule.

Example 4: A Limit of Type $0 \cdot \infty$

Find the limit
$$\lim_{x \to 0^+} x \ln x$$

Strategy. Since $\ln 0$ is undefined, but approaches $-\infty$, we have the form $0 \cdot (-\infty)$. Note, as $x \to 0$ then $y \to -\infty$. This one is tricky. The expression $x \ln x$ must be written as a quotient. x can be written as $\dfrac{1}{\frac{1}{x}}$. It's the reverse process of inverting the denominator and multiplying. Hence, we will rewrite $x \ln x$ as $\dfrac{\ln x}{\frac{1}{x}}$.

Solution. This limit has indeterminate form $0 \cdot (-\infty)$. We rewrite the product as a quotient.

$$\lim_{x \to 0^+} x \ln x = \lim_{x \to 0^+} \frac{\ln x}{\frac{1}{x}}$$

This limit has indeterminate form $\dfrac{-\infty}{\infty}$. Using L'Hôpital's rule and the following derivatives:

$$\frac{d}{dx}(\ln x) = \frac{1}{x} \text{ and } \frac{d}{dx}\left(\frac{1}{x}\right) = \frac{d}{dx}(x^{-1}) = -1 x^{-2} = \frac{-1}{x^2}$$

$$= \lim_{x \to 0^+} \frac{\frac{1}{x}}{\frac{1}{x^2}} = \lim_{x \to 0^+} \left(\frac{1}{x} \cdot \frac{-x^2}{1}\right) = \lim_{x \to 0^+} (-x) = 0$$

Summary. Using L'Hôpital's rule in this problem is a little harder than in the previous examples. It's somewhat tricky to rewrite the expression as a quotient. Once you complete that step correctly, the rest should follow relatively easily.

Example 5: A Limit of Type $\infty - \infty$

Find the limit.
$$\lim_{x \to 0} \left(\frac{1}{\sin x} - \frac{1}{x} \right)$$

Strategy. Since both terms will have 0 in the denominator if we try to substitute, neither will be defined. We will combine the two functions and then see if the criteria for L'Hôpital's rule is met.

Solution. This limit has indeterminate form $\infty - \infty$. We start by combining the fractions.

$$\lim_{x \to 0} \left(\frac{1}{\sin x} - \frac{1}{x} \right) = \lim_{x \to 0} \left(\frac{x - \sin x}{x \sin x} \right)$$

This in turn is a limit with indeterminate form $\frac{0}{0}$. Apply L'Hôpital's rule. Note that we need to use the product rule for the derivative of $x \sin x$ in the denominator.

$$= \lim_{x \to 0} \left(\frac{1 - \cos x}{\sin x + x \cos x} \right)$$

This is still a limit of type $\frac{0}{0}$. Apply L'Hôpital's rule once again. We have a product as part of the denominator again.

$$= \lim_{x \to 0} \left(\frac{\sin x}{\cos x + [\cos x + x(-\sin x)]} \right) = \lim_{x \to 0} \left(\frac{\sin x}{2 \cos x - x \sin x} \right)$$
$$= \frac{0}{2(1) - 0} = \frac{0}{2} = 0$$

Summary. Make sure that you understand all the steps in this problem. You need to understand *why* different steps are done as well as *how* they're done.

Example 6: A Limit of Type 1^∞

Find the limit.
$$\lim_{x \to 0^+} (1+x)^{\frac{1}{x}}$$

Strategy. As stated in the overview to this section, when there is an expression with a variable in the exponent, we need to take the natural logarithm (ln) and then use one of the properties of logarithms.

Solution. This indeterminate form is of type 1^∞. First rewrite the problem in the form of an equation.

$$y = \lim_{x \to 0} (1+x)^{\frac{1}{x}}$$

In order to separate the exponent from the base, we take the natural logarithm on both sides of the equation:

$$\ln y = \lim_{x \to 0^+} \ln(1+x)^{\frac{1}{x}}$$
$$= \lim_{x \to 0^+} \left[\frac{1}{x} \ln(1+x)\right]$$

This indeterminate form is of type $\infty \cdot 0$. We will rewrite this expression as a quotient, and then apply L'Hôpital's rule.

$$= \lim_{x \to 0^+} \frac{\ln(1+x)}{x} = \lim_{x \to 0^+} \frac{\frac{1}{1+x}}{1} = 1$$

A **very important** step here is to remember to undo the natural logarithm. We have:

$$\ln y = 1$$

which is the shorthand for $\log_e y = 1$. By the definition of logarithm we now have:

$$y = e^1 = e$$

Consequently, we can put the whole problem together and say:

$$\lim_{x \to 0^+} (1+x)^{\frac{1}{x}} = e$$

Summary. This may be the first time that you are using logarithms in a long time. Using it will help you make even more sense of logarithms and help you see times when they are quite helpful.

Example 7: A Limit of Type 0^0

Find the limit.
$$\lim_{x \to 0^+} x^x$$

Strategy. As in the previous example, we will rewrite the expression as an equation, and then take the natural logarithm on both sides. We also must remember to undo the logarithm as the last step in solving the problem.

Solution. This indeterminate form is of type 0^0. In order to separate the exponent from the base, we apply the natural logarithm to the equation

$$y = \lim_{x \to 0^+} x^x$$
$$\ln y = \lim_{x \to 0^+} \ln x^x$$
$$\ln y = \lim_{x \to 0^+} x \ln x$$

This indeterminate form is of type $0 \cdot (-\infty)$. This is now the same problem as in example 4. We will rewrite this expression as a quotient, apply L'Hôpital's rule, and simplify.

$$\lim_{x \to 0^+} \frac{\ln x}{\frac{1}{x}} = \lim_{x \to 0^+} \frac{\frac{1}{x}}{-\frac{1}{x^2}} = \lim_{x \to 0^+} \left(\frac{1}{x}\right)\left(-\frac{x^2}{1}\right) = \lim_{x \to 0^+} (-x) = 0$$

Next undo the natural log:

$$\ln y = 0 \implies y = e^0 = 1$$
$$y = 1 \implies \lim_{x \to 0^+} x^x = 1$$

Summary. Once again you have seen the use of logarithms as a tool which is often used along with L'Hôpital's Rule.

3.3. Maximum and Minimum Values

Key Topics.
- Local and Absolute Maximum and Minimum Values
- Critical Numbers and Fermat's Theorem
- The Closed Interval Method

Worked Examples.
- Locating Local and Absolute Maxima and Minima from Graph
- Another Example of Locating Extrema from Graph
- Finding Critical Points of a Function
- Finding Absolute Maximum and Minimum Values of a Continuous Function on a Closed Interval

Overview

The extreme values of a function are necessary in many instances. We are talking about the highest value of the function and the lowest value of the function. These will often be the *peaks* and *valleys* of the graph. We call these points the maximum and minimum values. We are usually interested in a region of the graph - not the entire graph. For example, if we know that we'll make the most profit by selling an infinite number of a product, that doesn't tell us anything.

Local and Absolute Maximum and Minimum Values. We need to know how much to produce with the constraints of time and money. Hence, we talk of the extreme values in a certain locality and they are appropriately called **local** extrema (extrema is plural and consists of both maximum and minimum values). Many books call these values **relative** extrema since they are the extrema relative to a certain interval. The text uses the word **local**.

If we know that the value is absolutely the highest or lowest point on a graph, then we can call the value an absolute maximum or an absolute minimum.

Critical Numbers and Fermat's Theorem. A number, c, in the domain of a function is called a critical number if either $f'(c) = 0$ or $f'(c)$ does not exist. Remember that the derivative gives the slope. So this tells us that a critical number is either a number where the derivative is zero which implies a horizontal tangent line, or it is a number where the derivative does not exist which implies either a vertical tangent, a discontinuity, or a corner point.

Fermat's Theorem tells us that if there is a local minimum then it **MUST** occur at a critical number.

The Closed Interval Method. If there exists a continuous function on a closed interval, then we are guaranteed at least one absolute maximum and at least one absolute minimum. Think about this - if you cut off a continuous graph so that you are looking at one section of it, including both endpoints, then there must be at least one highest point and at least one lowest point. If you have a graphing calculator it is like looking at your screen and considering only the part of the graph which you are seeing. The highest and lowest points will occur at either a critical number or at an endpoint. Hence, we will find the critical numbers in the interval and then evaluate the value of the graph at the critical numbers and at each endpoint. The highest value is the absolute maximum, and the lowest value is the absolute minimum.

Example 1: Locating Local and Absolute Maxima and Minima from Graph

Use the graph of the function $f(x) = (1-x)e^x$ to locate all local and absolute maximum and minimum values of $f(x)$ on the interval (a) $(-\infty, \infty)$, and (b) $[0, 1]$.

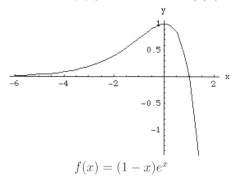

$f(x) = (1-x)e^x$

Strategy. For part a) we will look at the entire graph and see if there are highest and lowest points which can be determined.

For part b) we are guaranteed that there will be both an absolute maximum and an absolute minimum. We will look at the part of the graph on the interval [0,1] to determine these values.

Solution. a) On the interval $(-\infty, \infty)$, we can see from the graph that f has a local and absolute maximum value of 1 at $x = 0$. Since the graph on the left is approaching, but not touching the x-axis, f has neither a local nor a absolute minimum on the interval $(-\infty, \infty)$.

b) Since $[0, 1]$ is a closed interval, we are guaranteed both an absolute maximum and an absolute minimum. We can see from the graph that f has an absolute maximum value of 1 at $x = 0$, and an absolute minimum value of 0 at $x = 1$. However, f has no local maximum or minimum values on the given interval.

Summary. We can see the maximum and minimum from looking at the graph. The absolute maximum and the absolute minimum are the absolutely highest and lowest values reached. The local maximum and the local minimum will often look like a peak or valley of the graph and hence, can never occur at an endpoint of an interval.

Example 2: Another Example of Locating Extrema from Graph

Use the graph of the function f to locate all local and absolute maximum and minimum values of f on the interval a) $(-1, 3)$, and b) $[-1, 3]$.

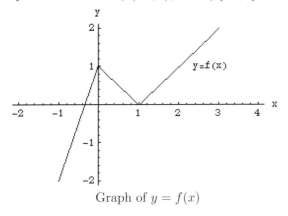

Graph of $y = f(x)$

Strategy. For part a), we have an open interval. There may or may not be extrema on this interval. However, part b) has a closed interval and, therefore, we are guaranteed both an absolute maximum and an absolute minimum.

Solution. a) On the open interval $(-1, 3)$, we can see from the graph that f has a local maximum value of 1 at $x = 0$, and a local minimum value of 0 at $x = 1$. f has no absolute maximum or minimum values on this interval.

b) On the closed interval $[-1, 3]$, we can see from the graph that f has a local maximum value of 1 at $x = 0$, and a local minimum value of 0 at $x = 1$. f also has an absolute maximum value of 2 at $x = 3$, and an absolute minimum value of -2 at $x = -1$.

Summary. As you can see, it is necessary to first see if you have an open or closed interval and then simply identify the highest and lowest points and where they occur.

Example 3: Finding Critical Numbers of a Function

Find all critical numbers of the function $f(x) = \sqrt{1-x^2}$ on the interval $[-1, 1]$.

Strategy. Critical numbers occur at x values in the domain of the function where the derivative is zero or the derivative does not exist. We will find the derivative and then determine where it is equal to zero and where it doesn't exist.

Solution.
$$f(x) = \sqrt{1-x^2} = (1-x^2)^{\frac{1}{2}}$$
Using the power rule and the chain rule, we get
$$f'(x) = \frac{1}{2}(1-x^2)^{\frac{-1}{2}}(-2x) = \frac{-2x}{2\sqrt{1-x^2}} = \frac{-x}{\sqrt{1-x^2}}$$

Critical numbers are numbers where $f' = 0$, or where f' is undefined. Here $f' = 0$ at $x = 0$, and f' is undefined at $x = 1$ and $x = -1$. Hence, the critical numbers are $x = 0$, $x = 1$, and $x = -1$.

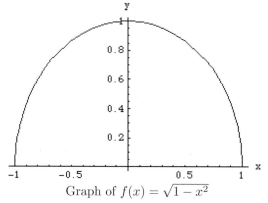

Graph of $f(x) = \sqrt{1-x^2}$

From the graph, we can see that f has a local and absolute maximum value of 1 at $x = 0$, and an absolute minimum value of 0 at both $x = -1$ and $x = 1$.

Summary. When you are looking at the point where the extrema exists, remember that the x-coordinate is the critical number and that the y-coordinate is the maximum or minimum value. In a later section we will learn how to find these values without looking at a graph.

Example 4: Finding Absolute Maximum and Minimum Values of a Continuous Function on a Closed Interval

Find the absolute maximum and minimum values of the function $f(x) = x^3 - 3x + 5$ on the interval $[0, 2]$.

Strategy. Here we have a continuous function on a closed interval. Hence, we will find any critical numbers on the interval and then calculate the values of the function at the critical numbers and at each endpoint to find the highest and lowest values. We are guaranteed answers since we are on a closed interval.

Solution. $f'(x) = 3x^2 - 3$. We need to find x values which make $f'(x) = 0$.

$$\begin{align} 3x^2 - 3 &= 0 \\ x^2 - 1 &= 0 \\ (x-1)(x+1) &= 0 \\ &\implies x = 1, x = -1 \end{align}$$

Since $x = -1$ is not in the interval $[0, 2]$ the only critical number which we use is $x = 1$. We need to compare the values of $f(x)$ at the critical number and at the endpoints.

x	$f(x)$
0	5
1	3 = absolute minimum value
2	7 = absolute maximum value

The absolute minimum value of f is 3 and occurs at $x = 1$, and the absolute maximum value of f is 7 and occurs at $x = 2$.

Summary. This problem was a great example of finding extrema on a closed interval. We had two values for $f'(x) = 0$. Only one value was in the given interval. We checked the value of $f(x)$ at the critical number and at each endpoint. Then we call the largest value the absolute maximum, and the smallest value the absolute minimum.

3.4. Increasing And Decreasing Functions

Key Topics.
- Increasing and Decreasing Functions And the First Derivative
- The First Derivative Test
- Using the First Derivative to Sketch Graphs

Worked Examples.
- Using the First Derivative to Sketch the Graph of a Polynomial
- Using the First Derivative to Sketch the Graph of a Rational Function
- Sketching a Graph with Given Properties

Overview

Beginning with this section, we will see some of the purposes of a derivative. The main idea to keep in mind is that the derivative measures change and gives us the slope of a graph.

Increasing and Decreasing Functions And the First Derivative. Since the derivative gives us the slope, we can tell if a function is increasing or decreasing on an interval just by knowing the sign of the first derivative at any point in the interval. Pretty amazing, isn't it? If the derivative is positive, then the slope is positive and the function is increasing. If the derivative is negative, then the slope is negative and the function is decreasing.

The First Derivative Test. For the First Derivative Test we must find the critical numbers and place them on a number line. Then we can see whether the function is increasing or decreasing on both the right and left sides of each critical number. If the function goes from increasing on the left to decreasing on the right ($f'(x) > 0$ to $f'(x) < 0$), then the graph is going up and then down giving us a local maximum. Try this shape with your hand - go up and then down - and you will see that you are forming a maximum.

Similarly, if the function goes from decreasing to increasing ($f'(x) < 0$ to $f'(x) > 0$) - down and then up - we have a local minimum.

Using the First Derivative to Sketch Graphs. Once we know where any local maximum and local minimum are, we can plot those points. It is often relatively easy to also find the y-intercept. Then we can connect all the dots to form the graph of the function.

Example 1: Using the First Derivative to Sketch the Graph of a Polynomial

Find the intervals where the function $f(x) = x^4 - 8x^2 + 8$ is increasing and decreasing, and use this to sketch the graph of $f(x)$.

Strategy. First we will find $f'(x)$. Secondly, we will note that $f'(x)$ is everywhere defined, so we will find critical numbers by solving the equation $f'(x) = 0$.

Thirdly, we will place all the critical number on a number line.

Lastly, we will calculate $f'(x)$ on each interval we have formed to determine whether the derivative is positive or negative on that interval. This will tell us whether the function is increasing or decreasing on each interval.

After we know where the function is increasing and where the function is decreasing we can determine if we have any local extrema using the First Derivative Test.

Solution. We first find the derivative, and all critical numbers.

$$\begin{aligned} f(x) &= x^4 - 8x^2 + 8 \\ f'(x) &= 4x^3 - 16x = 4x(x^2 - 4) = 4x(x+2)(x-2) \end{aligned}$$

To get the critical numbers, we need to solve $f'(x) = 4x(x+2)(x-2) = 0$. From this we find that the critical numbers are $x = 0$, $x = -2$, and $x = 2$. Draw a number line and put marks at these three numbers.

$$\underset{x < -2 \quad \mid \quad -2 < x < 0 \quad \mid \quad 0 < x < 2 \quad \mid \quad x > 2}{\rule{8cm}{0.4pt}}$$

We need to see if the derivative is positive or negative on each of these four intervals we formed.

If $x < -2$, then all three factors of f', that is $4x, x+2$, and $x-2$, are negative. (You can see this by choosing any number in the interval $x < -2$. For example if you choose $x = -3$, then you have the three factors: $4(-3) < 0$, $(-3+2) < 0$ and $(-3-2) < 0$. Three negative factors make the product negative. You can check all the intervals by choosing a test number in the interval if it helps you to better understand what is happening.)

Therefore $f'(x) = 4x(x+2)(x-2)$ is negative, and the function $f(x)$ is decreasing for $x < -2$.

In a similar manner if $-2 < x < 0$, then the factor $x + 2$ is positive, and the other two factors are negative, therefore $f'(x)$ is positive on this interval. Thus $f(x)$ is increasing on the interval $(-2, 0)$.

Also, if $0 < x < 2$, then the factor $x - 2$ is negative, and the other two factors are positive, therefore $f'(x)$ is negative on this interval. Thus $f(x)$ is decreasing on the interval $(0, 2)$.

And if $x > 2$, then all three factors of f' are positive, and therefore $f'(x)$ is positive. Thus the function $f(x)$ is increasing for $x > 2$.

Altogether we have, f is increasing on the intervals $(-2, 0)$, and $(2, \infty)$, and f is decreasing on the intervals $(-\infty, -2)$, and $(0, 2)$.

We can organize the work we just did in the following diagram.

sign of $4x$	$-$	$-$	$+$	$+$
sign of $x+2$	$-$	$-$	$-$	$+$
sign of $x-2$	$-$	$+$	$+$	$+$
sign of f'	$-$	$+$	$-$	$+$
intervals	$x<-2$	$-2<x<0$	$0<x<2$	$x>2$
f increasing /decreasing	↘	↗	↘	↗

From the diagram, we can see that we have a local minimum at $x = -2$, a local maximum at $x = 0$, and a local minimum at $x = 2$. To plot the points, we need both the x and y-coordinates. Since we started with $f(x) = x^4 - 8x^2 + 8$, there is a local minimum at $(-2, -8)$ and at $(2, -8)$. The local maximum is at $(0, 8)$ which is also the y-intercept. Now we can sketch the graph of $f(x)$.

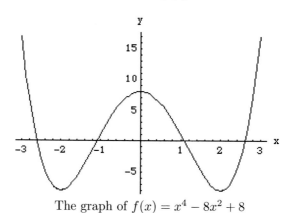

The graph of $f(x) = x^4 - 8x^2 + 8$

Summary. If you have a graphing calculator, you can graph this function is the standard window and check to see that all the extrema we found are correct.

Example 2: Using the First Derivative to Sketch the Graph of a Rational Function

$$f(x) = \frac{2}{4 - x^2}$$

Find all asymptotes of this function, and the intervals where the function is increasing and decreasing, and use this information to sketch the graph of $f(x)$.

Strategy. Let's have a short review on how to find vertical and horizontal asymptotes in an algebraic rational function.

The easier ones to find are the vertical asymptotes. These occur wherever the denominator is equal to zero after any reducing has been done. In this function, the denominator is $4 - x^2$. To find vertical asymptotes, set this equal to zero and solve.

$$4 - x^2 = 0$$
$$(2 - x)(2 + x) = 0$$
$$x = 2 \text{ or } x = -2$$

Hence the two vertical asymptotes are at $x = 2$ and at $x = -2$.

The horizontal asymptote is trickier. One way to find it is to find the limit of the function as $x-> \pm\infty$. Or you can just remember the results of that which are:

a) If the degree of the numerator is less than the degree of the denominator, then $y = 0$ is the horizontal asymptote.

b) If the degree of the numerator is equal to the degree of the denominator, the $y = $ the ratio of the leading coefficients is the horizontal asymptote. For example, is the problem looks like $y = \dfrac{4x^3 + ...}{3x^3 + ...}$ then $y = \dfrac{4}{3}$ is the horizontal asymptote.

c) In any other case, there is no horizontal asymptote.

In our problem, the degree of the numerator is 0, which is less than the degree of the denominator which is 2. Hence, $y = 0$ is the horizontal asymptote.

Solution. As explained above, this function has two vertical asymptotes, $x = -2$, and $x = 2$.

Also, since $\lim\limits_{x \to \pm\infty} \dfrac{2}{4 - x^2} = 0$, and as explained above, the degree of the numerator is less than the degree of the denominator, f has a horizontal asymptote at $y = 0$, which is the x-axis.

We can easily get the y-intercept which is $(0, \frac{1}{2})$.

To find the local maximum and the local minimum we need to find the first derivative using the quotient rule.

$$f'(x) = \frac{0 \cdot (x^2 - 4) - 2 \cdot (-2x)}{(4 - x^2)^2} = \frac{4x}{(4 - x^2)^2}$$

The critical numbers occur at any numbers in the domain where $f'(x)$ is either equal to zero or does not exist. The only critical number here is $x = 0$. (± 2 are not critical numbers because they are not in the domain of the function.)

When we draw the number line, we **must** put the critical number ($x = 0$) on it, and we must also identify where the vertical asymptotes are, so that we can see what the function is doing on either side of the asymptotes.

$$\begin{array}{c|c|c|c} x<-2 & -2<x<0 & 0<x<2 & x>2 \end{array}$$

Now we need to see if the derivative is positive or negative on each of these four intervals we formed. Note that $\left(4 - x^2\right)^2$ is always greater than or equal to zero. Therefore the sign of $f'(x)$ only depends on the factor $4x$.

If $x < -2$, then $4x$ is negative, and therefore $f'(x) = \dfrac{4x}{\left(4 - x^2\right)^2}$ is negative, and the function $f(x)$ is decreasing for $x < -2$.

If $-2 < x < 0$, then the factor $4x$ is still negative, therefore $f'(x)$ is negative on this interval. Thus $f(x)$ is decreasing on the interval $(-2, 0)$.

If $0 < x < 2$, then the factor $4x$ is positive, therefore $f'(x)$ is positive on this interval. Thus $f(x)$ is increasing on the interval $(0, 2)$.

If $x > 2$, then $4x$ is positive, and therefore $f'(x)$ is positive. Thus the function $f(x)$ is increasing for $x > 2$.

Altogether we have that f is decreasing on the intervals $(-\infty, -2)$, and $(-2, 0)$, and that f is increasing on the intervals $(0, 2)$, and $(2, \infty)$.

We can organize the work again in a diagram.

sign of $4x$	$-$	$-$	$+$	$+$
sign of $\left(4 - x^2\right)^2$	$+$	$+$	$+$	$+$
sign of f'	$-$	$-$	$+$	$+$
intervals	$x < -2$	$-2 < x < 0$	$0 < x < 2$	$x > 2$
f increasing /decreasing	↘	↘	↗	↗

From the diagram, we can see that we have a local minimum at $x = 0$.

Combining the information about asymptotes, the y-intercept and increasing and decreasing, we can now sketch the graph of $f(x)$.

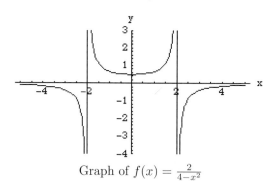

Graph of $f(x) = \frac{2}{4-x^2}$

Summary. If you think of all these parts as putting together a puzzle, it will be easier for you. In the next section, we will add the last piece of the puzzle when we talk about concavity and the second derivative.

Example 3: Sketching a Graph with Given Properties

Sketch a graph with the following properties.

$$f'(x) > 0 \text{ for } x < 0, \quad f'(x) < 0 \text{ for } x > 0$$
$$\lim_{x \to -\infty} f(x) = -\infty, \quad \lim_{x \to +\infty} f(x) = 1$$
$$f(0) = 2$$

Strategy. When we see $f'(x) > 0$ for $x < 0$, we need to read that the derivative is positive for $x < 0$. In other words, the graph is increasing for all $x < 0$.

Similarly, $f'(x) < 0$ for $x > 0$ says that the derivative is negative for $x > 0$. In other words, the graph is decreasing for all $x > 0$.

Solution. From the strategy above, we know that f is increasing to the left of $x = 0$, and decreasing to the right of $x = 0$. At $x = 0$ there is a local maximum. Since we are also told that $f(0) = 2$, we know that the local maximum is at $(0, 2)$.

The two statements involving limits give us the horizontal asymptote. They tell us what is happening to the graph as we get far away from $(0, 0)$.

On the left, the graph just goes to $-\infty$ and there is no horizontal asymptote.

But on the right, as x gets very large, y or $f(x)$ approaches 1. Therefore, there is a horizontal asymptote at $y = 1$ on the right side of the graph.

Putting all of this information together, we get a graph similar to this one:

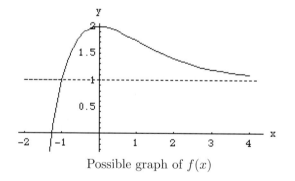

Possible graph of $f(x)$

Summary. It helps a lot to read $f'(x) > 0$ as the derivative is positive and $f'(x) < 0$ as the derivative is negative. In other words, read the meaning of the symbols in words to help sketch the graph.

3.5. Concavity And Overview Of Curve Sketching

Key Topics.
- Concavity And the Second Derivative
- The Second Derivative Test
- Using the First and Second Derivative to Sketch Graphs

Worked Examples.
- Using the First and Second Derivative to Sketch the Graph of a Polynomial
- Using the First and Second Derivative to Sketch the Graph of a Rational Function
- Using the Second Derivative Test

Overview

We now know how to find where a function is increasing and where a function is decreasing. We can also find local extrema. We have one more task before we can completely sketch the graph. We need to know *how* it is increasing or decreasing. Does it *bend down* or does it *bend up*. Here we are getting into the subtleties of the graph. We can do this using the second derivative.

Concavity And the Second Derivative

Since a derivative measures how a function changes, the 2nd derivative measures how the 1st derivative changes. If the 2nd derivative is positive, then the slope (1st derivative) is increasing making the graph *bend upwards*. We call this concave up. (Think of the graph of $y = x^2$.)

If the 2nd derivative is negative, then the slope (1st derivative) is decreasing making the graph *bend down*. We call this concave down. (Think of the graph of $y = -x^2$).

A poem to help you remember:

    Concave up looks like a cup

    Concave down looks like a frown.

A bit silly maybe, but it will help you remember. The point where the concavity changes is called a **point of inflection** or an **inflection point.**

The Second Derivative Test

Note that the full name here is **The Second Derivative Test for Extrema**. We use the 2nd derivative for two different processes which are related but not the same. DO NOT confuse them. Above we used the 2nd derivative to find concavity and points of inflection. We have a separate process here. This is another way to use the 2nd derivative. As the full name states, we will find the extrema using the 2nd derivative.

If we get extremely logical, we can put what we know about critical numbers together with what we know about concavity. If there is a critical number where the 1st derivative is equal to zero, then there may be a local maximum or a local minimum - a *peak* or a *valley* - at that point. If we test the second derivative **at the critical number,** CN, and find $f''(CN) > 0$, then we know the graph is concave up meaning you have a *valley*. Hence, at that critical number there is a local minimum.

Similarly, if $f''(CN) < 0$, then we know the graph is concave down, meaning we have a *peak*. Hence, at that critical number, there is a local maximum.

In symbols, if CN is a critical number and:

$$f''(CN) > 0, \text{ then there is a local minimum at } CN$$
$$f''(CN) < 0, \text{ then there is a local maximum at } CN$$

Using the First and Second Derivative to Sketch Graphs

Here we will just add to what we did in the previous section. We will first use the 1st derivative to see where the function is increasing and decreasing and find local extrema.

Next, we will find the 2nd derivative and see where it is zero or undefined. These values will be possible points of inflection. Then we can check the concavity on the intervals surrounding the possible points of inflection and determine if there is a change in concavity. If there is, then point is a point of inflection.

The definition of a point of inflection is a point in the domain where concavity changes.

Note: The process for finding points of inflection using the 2nd derivative exactly parallels the process for finding local extrema using the 1st derivative.

Example 1: Using the First and Second Derivative to Sketch the Graph of a Polynomial

Sketch the graph of the function $f(x) = \frac{1}{3}x^3 - x^2 - 3x + 3$.

Strategy. In this problem, we will use the 1st derivative and find the extrema. Then we will repeat the process using the 2nd derivative to find concavity and any points of inflection. Lastly, we will sketch the graph by putting together all the information we gathered.

Solution. We first find the first and second derivatives.

$$\begin{aligned} f(x) &= \frac{1}{3}x^3 - x^2 - 3x + 3 \\ f'(x) &= x^2 - 2x - 3 = (x+1)(x-3) \\ f''(x) &= 2x - 2 \end{aligned}$$

Setting $f'(x) = 0$ in order to find the critical numbers we get $(x+1)(x-3) = 0$. Therefore, the critical numbers are $x = -1$, and $x = 3$. Draw number lines for each factor of f', and put marks at $x = -1$, and $x = 3$.

sign of $x - 3$	−	−	+
sign of $x + 1$	−	+	+
sign of $f'(x)$	+	−	+
interval	$x < -1$	$-1 < x < 3$	$x > 3$
f increasing/decreasing	↗	↘	↗

The function f is increasing on the intervals $(-\infty, -1)$ and $(3, \infty)$, and f is decreasing on the interval $(-1, 3)$. At $x = -1$, f has a local maximum, and at $x = 3$, f has a local minimum. Using the original function to obtain the y-values, we now have a local maximum at $(-1, \frac{14}{3})$ and a local minimum at $(3, -6)$.

We use the second derivative to determine the concavity. We will determine where f'' is zero or undefined. We need to solve $f''(x) = 2x - 2 = 0$, which leads us to the conclusion that f'' has only one zero at $x = 1$. Draw a number line for f'', and put a mark at $x = 1$, and organize the data for $f''(x)$.

sign of $f''(x)$	−	+
interval	$x < 1$	$x > 1$
f concave up/down	∩	∪

The graph of f is concave down for $x < 1$, and concave up for $x > 1$. Since the concavity changes at $x = 1$, there is a point of inflection at $x = 1$.

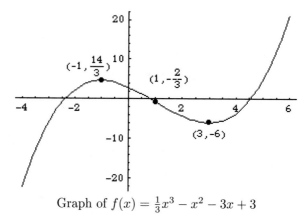

Graph of $f(x) = \frac{1}{3}x^3 - x^2 - 3x + 3$

Summary. Because the function was a polynomial, we didn't have to worry about vertical or horizontal asymptotes. Also, both derivatives were factorable which made the process of obtaining the critical numbers and point of inflection straightforward. Actually, quite a nice problem!

Example 2: Using the First and Second Derivative to Sketch the Graph of a Rational Function

Sketch the graph of the function

$$f(x) = \frac{x^2}{(x-2)^2}$$

Strategy. Since this is a rational function, we do need to look for asymptotes. The denominator is equal to zero when $x = 2$ making $x = 2$ a vertical asymptote.

When the denominator is multiplied out, the function becomes $f(x) = \frac{x^2}{x^2 - 4x + 4}$ which lets us see that the degree of the top is equal to the degree of the bottom. Hence, the horizontal asymptote is the ratio of the two leading coefficients: $y = \frac{1}{1} = 1$. For an explanation of this, see section 3.4 example #2 in this book.

Next, we will find the 1st and 2nd derivatives to find critical numbers, local extrema, concavity, and points of inflection.

Solution. From the strategy section above, we know that the vertical asymptote is at $x = 2$, and that the horizontal asymptote is at $y = 1$.

(Another way to obtain the horizontal asymptote is to say that since $\lim\limits_{x \to \pm\infty} \frac{x^2}{(x-2)^2} = 1$, f has a horizontal asymptote $y = 1$.)

Again, we find both derivatives.

$$f(x) = \frac{x^2}{(x-2)^2}$$

$$f'(x) = \frac{2x(x-2)^2 - x^2(2(x-2))}{(x-2)^4} \quad \text{(factor out } 2(x-2)\text{)}$$

$$= \frac{2(x-2)[x(x-2) - x^2]}{(x-2)^4} = \frac{2[x^2 - 2x - x^2]}{(x-2)^3}$$

$$= \frac{2x^2 - 4x - 2x^2}{(x-2)^3} = \frac{-4x}{(x-2)^3}$$

$$f''(x) = \frac{-4(x-2)^3 - (-4x)\left(3(x-2)^2\right)}{(x-2)^6} = \frac{(x-2)^2[-4(x-2) + 12x]}{(x-2)^6}$$

$$= \frac{-4(x-2) + 12x}{(x-2)^4}$$

$$= \frac{-4x + 8 + 12x}{(x-2)^4} = \frac{8x + 8}{(x-2)^4}$$

Setting the 1st derivative equal to zero, we find the only critical number is $x = 0$. The derivative does not exist at $x = 2$, but this is not a critical number since it is not in the domain of the original function. Remember that $x = 2$ is the vertical asymptote. When you draw the number lines, you must also put marks where the vertical asymptote is so that you can see what the function is doing on either side of it. Draw number lines for each factor of f', and put marks at $x = 0$, and $x = 2$.

sign of $-4x$	+	−	−
sign of $(x-2)^3$	−	−	+
sign of $f'(x)$	−	+	−
intervals	$x < 0$	$0 < x < 2$	$x > 2$
f increasing/decreasing	↘	↗	↘

The function f is increasing on the interval $(0, 2)$, and f is decreasing on the intervals $(-\infty, 0)$ and $(2, \infty)$. At $x = 0$, f has a local minimum, and at $x = 2$, f has a vertical asymptote.

We use the second derivative to determine the concavity. We need to see where f'' is zero or undefined but in the domain of the function. $f'' = 0$ when $x = -1$, and f'' is undefined when $x = 2$ which cannot be a point of inflection since it is not in the domain of the function. We will still place it on the number line so that we can check the concavity on either side of it. Draw a number line for each factor of f'', and put marks at $x = -1$, and at $x = 2$.

sign of $8x + 8$	−	+	+
sign of $(x-2)^4$	+	+	+
sign of $f''(x)$	−	+	+
intervals	$x < -1$	$-1 < x < 2$	$x > 2$
f concave up/down	∩	∪	∪

f is concave down for $x < -1$, and concave up for $-1 < x < 2$, and $x > 2$. At $x = -1$, f has a point of inflection because the concavity changes on either side of it.

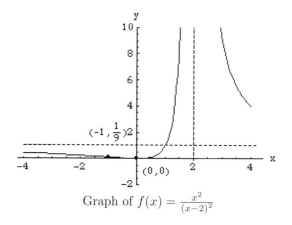

Graph of $f(x) = \frac{x^2}{(x-2)^2}$

Summary. There was a lot to do in this problem. It is important to go very logically step-by-step.
1) Find the y-intercept
2) Find any asymptotes
3) Find $f'(x)$ and $f''(x)$
4) Use $f'(x)$ to find the critical numbers

5) Use the critical numbers to find the intervals of increasing and decreasing and local extrema
6) Use $f''(x)$ to find possible points of inflection
7) Use the possible points of inflection to find the concavity on the intervals.
8) If there is a change in concavity, then identify the point of inflection.
9) Put all the information together and sketch the graph.

Example 3: Using the Second Derivative Test

Find all local maxima and minima of the function $f(x) = x \ln x$ for $x > 0$.

Strategy. Since the 2nd derivative is easily found for this function, we will use the **Second Derivative Test for Extrema** here. First we will find the critical numbers and then test the critical numbers in the 2nd derivative to determine concavity. Putting the two facts together, we will know whether the critical number gives a local maximum or a local minimum.

Note that if the 2nd derivative at the critical number is equal to zero, then there is no conclusion. We would need to go back and use the First Derivative Test to determine if there is a local minimum, a local maximum, or neither.

Solution. First we use the product rule to find the first derivative in order to locate all critical numbers.

$$f'(x) = \ln x + x \cdot \frac{1}{x} = \ln x + 1$$

The critical numbers occur where $f' = 0$, or f' is undefined. $f'(x)$ is defined for all $x > 0$, thus we only need to find the points where $f' = 0$.

$$\begin{aligned} f'(x) &= 0 \\ \ln x + 1 &= 0 \\ \ln x &= -1 \\ e^{-1} &= x \implies x = e^{-1} = \frac{1}{e} \end{aligned}$$

f has only one critical point, $x = \frac{1}{e}$.

The second derivative is $f''(x) = \frac{1}{x}$, and

$$f''(\frac{1}{e}) = \frac{1}{\frac{1}{e}} = e > 0$$

Since the second derivative at the critical point $x = \frac{1}{e}$ is positive, the graph is concave up at that point and hence, there is a local minimum at $x = \frac{1}{e}$. The local minimum is at

$$(\frac{1}{e}, f(\frac{1}{e})) = (\frac{1}{e}, \frac{1}{e} \ln \frac{1}{e}) = (\frac{1}{e}, \frac{-1}{e})$$

(Note: $\ln \frac{1}{e} = \ln e^{-1} = -1 \ln e = -1$)

Summary. When we want to know if a critical number is a local max or a local min and the 2nd derivative is easy to find, we may want to use the Second Derivative Test for Extrema. Please remember, however, that we can always use the First Derivative Test with the number line to determine if a critical number is a local max, a local min, or neither.

3.6. Optimization

Key Topics.
- Optimization

Worked Examples.
- A Fence Problem
- Finding Shortest Distance
- An Optimization Problem
- A Cylindrical Box Problem

Overview

As you can imagine, we often need to find the largest or the smallest of something. For example, if you own a company, you need the largest profit or the smallest cost. You might want to maximize the amount of cola that will fit in a can with a certain shape. Or maximize the area of a garden when you have a fixed amount of fencing. All of these are called optimization problems. To optimize something means to find the largest and/or the smallest value. In other words, to find the maximum and/or minimum values.

Optimization. The very first thing to do in these problems is to find the equation or formula for the quantity that needs to be optimized. Then, just as we did in the previous section, we need to find the critical numbers by finding the derivative and seeing where it is equal to zero. If we have a closed interval to work with, we then need to calculate the function at any critical numbers in the interval and at the endpoints of the interval to identify the maximum and/or the minimum.

Remember: Find a formula for the function of the quantity that we have to optimize, then take the derivative of that function. For example, if you are looking for maximum volume, find V', for minimum cost, find C', for maximum height, find h' etc. Often the formula has more than one variable. In that case there will be other information in the problem relating the two variables allowing you to use substitution so that there is only one variable when you take the derivative.

3.6. OPTIMIZATION

Example 1: A Fence Problem

A three-sided fence is to be built next to a straight wall, which forms the fourth side of a rectangular region. The enclosed area is to equal $1800\,\text{ft}^2$. Find the dimensions that will minimize the length of the fence.

Strategy. This is a typical optimization problem. In a set of optimization problems there is usually at least one *fence* problem where one side doesn't need any fencing. We absolutely must sketch a picture and label the sides. In this problem, let's call the sides perpendicular to the wall x, and the side parallel to the wall y.

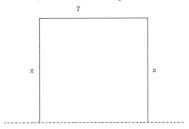

Solution. Then the area is $A = xy$ since there is a rectangular region. We don't exactly have the perimeter since we are not going all the way around, so the amount of fencing needed is $2x + y$. Let's call that F. So, $F = 2x + y$ is the amount of fencing, which is the quantity to be minimized.

But, there are two variables. We know one other bit of information, which is that the area is $1800\,\text{ft}^2$. Consequently, we have:

$$\begin{aligned}\text{Area} &= xy = 1800\,\text{ft}^2 \\ y &= \frac{1800}{x}\end{aligned}$$

Hence we can write:

$$\begin{aligned}F &= 2x + y \\ &= 2x + \frac{1800}{x} \\ &= 2x + 1800x^{-1}\end{aligned}$$

It is very important to specify the domain. Here the allowable values for x are $x > 0$.

To find the minimum, we must first find the critical numbers.

$$\begin{aligned}F'(x) &= 2 - 1800x^{-2} \\ &= 2 - \frac{1800}{x^2}\end{aligned}$$

$F'(x)$ is undefined for $x = 0$, but $x = 0$ is not in the domain. We need to solve for $F'(x) = 0$:

$$\begin{aligned} 2 &= \frac{1800}{x^2} \\ 2x^2 &= 1800 \\ x^2 &= 900 \\ x &= \pm 30 \end{aligned}$$

Since the domain is restricted to $x > 0$, we only get one critical point $x = 30$. We will make sure that this is a minimum exactly as we did in the last section. Draw a number line and mark the critical point.

$$\underset{0 < x < 30 \quad \mid \quad x > 30}{\rule{3cm}{0.4pt}\vert\rule{3cm}{0.4pt}}$$

We need to check the two intervals to see if the function is increasing or decreasing on them.

When $0 < x < 30$, $F'(x) < 0$. Hence, the function is decreasing on that interval.

When $x > 30$, $F'(x) > 0$ and the function is increasing on that interval. Therefore we know that the function has an absolute minimum at $x = 30$.

sign of $F'(x)$	$-$	$+$
intervals	$0 < x < 30$	$x > 30$
f increasing/decreasing	↘	↗

Hence, the dimensions of the fence are $x = 30$ feet, and $y = 60$ feet.

Summary. If you identify the variables and identify the function which needs to be optimized, then these problems are not so difficult. Also, do not forget to answer the question including the units needed.

Example 2: Finding Shortest Distance

Find the shortest distance from the point $(4,0)$ to a point on the parabola $y^2 = 2x$.

Strategy. You have been doing distance problems for several years. The distance formula follows directly from the Pythagorean Theorem. The distance, d, between two points (x_1, y_1) and (x_2, y_2) is:

$$(d)^2 = (x_2 - x_1)^2 + (y_2 - y_1)^2$$
$$d = \sqrt{(x_2 - x_1)^2 + (y_2 - y_1)^2}$$

Any point on the parabola has coordinates (x, y) where $y^2 = 2x$. The other point is given as $(4, 0)$.

Solution. We first draw a picture.

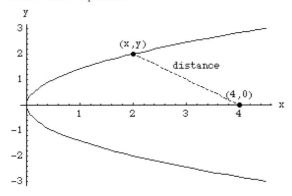

Our goal is to minimize the distance between a point (x, y) on the parabola and the point $(4, 0)$. It is easier to minimize the distance squared.

$$d^2 = (x - 4)^2 + (y - 0)^2$$
$$= x^2 - 8x + 16 + y^2$$

Since $y^2 = 2x$ we substitute $2x$ for y^2 and combine like terms. The function we want to minimize is

$$f(x) = x^2 - 8x + 16 + 2x$$
$$= x^2 - 6x + 16$$

The domain of f is $[0, \infty)$.

Now $f'(x) = 2x - 6$, so the only critical number is $x = 3$. As $f'(x) < 0$ for all $x < 3$ and $f'(x) > 0$ for all $x > 3$, we know that the absolute minimum occurs when $x = 3$. To find y, we go back to:

$$y^2 = 2x$$
$$y^2 = 2(3)$$

$$y^2 = 6$$
$$y = \pm\sqrt{6}$$

Hence the points $(3, \sqrt{6})$ and $(3, -\sqrt{6})$ are the closest points to $(4,0)$. The shortest distance, then, from the point $(4,0)$ to the parabola is:

$$\begin{aligned} d &= \sqrt{(4-3)^2 + (0-\sqrt{6})^2} \\ &= \sqrt{1+6} = \sqrt{7} \end{aligned}$$

Summary. What's new here is expressing any point on the parabola as (x, y). The rest is simply the distance formula which you learned a long time ago.

3.6. OPTIMIZATION

Example 3: An Optimization Problem

A rectangular building is to cover 20,000 square feet. Zoning regulations require 20 feet of space at the front and rear, and 10 feet of space on either side. Find the dimensions of the smallest rectangular piece of property on which the building can be legally constructed.

Strategy. This problem requires a very careful reading. It also needs a sketch to show the placement of the building and the surrounding space. If L and W represent the length and width of the building, then $L + 20$ and $W + 40$ represent the length and width of the entire piece of property. Once that is found, the problem is more easily solved.

Solution.

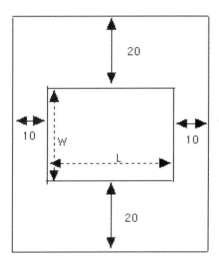

The 20,000 square feet represents the area of the building alone. Since the dimensions of the building are L and W we can write:

$$A = L \cdot W$$
$$20000 = L \cdot W$$
$$\frac{20000}{W} = L$$

The area of the building plus the "extra" space is then $A = (L + 20)(W + 40)$. Substituting $\frac{20000}{W}$ for L we have:

$$\begin{aligned} A &= \left(\frac{20000}{W} + 20\right)(W + 40) \\ &= 20,000 + \frac{800000}{W} + 20W + 800 \\ &= 20,000 + 800,000 W^{-1} + 20W + 800 \end{aligned}$$

$$A' = -800,000W^{-2} + 20$$

Now, $A' = 0$ implies that

$$0 = \frac{-800,000}{W^2} + 20$$
$$\frac{800,000}{W^2} = 20$$
$$800,000 = 20W^2$$
$$40000 = W^2$$

$$\implies W = 200 \text{ and } L = \frac{20,000}{200} = 100$$

Hence, the dimensions of the region are: Width $= W + 40 = 240$ feet, and length $= L + 20 = 120$ feet.

Summary. This problem definitely requires a picture and an extremely careful reading so that we understand what is being asked and what is given.

Example 4: A Cylindrical Box Problem

Suppose we have to make a cylindrical box of volume $25\,\text{ft}^3$. The top and the bottom circles are to be made of cardboard which costs \$0.50 per ft^2, and the side is made by bending a rectangular piece of metal sheet which costs \$5 per ft^2. What should the dimensions of the cylinder be if we want to minimize the cost?

Strategy. Once again a picture is essential. The side is a rectangle if we open it up. We need to find the area of the two circles and the area of the side, and then multiply each by its respective cost. Then we will minimize the cost.

Solution. Draw a picture of the cylinder, and a picture of the unfolded cylinder, and label all sides.

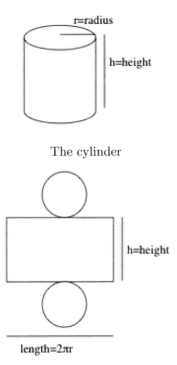

The cylinder

The cylinder *unfolded*

The area of a circle is: $A = \pi r^2$. The cost is \$0.50 per ft^2, so the cost of the top and the bottom (two circles) is $C_1 = .50(2\pi r^2) = \pi r^2$.

The area of the rectangular sheet is length times width, which in this case is $2\pi r h$. You can understand the $2\pi r$ if you think of *unfolding* the side of the cylinder. That length is the circumference of the circular bottom and circumference is equal to $2\pi r$. The side costs \$5 per square foot and hence the cost of the side is $C_2 = 5(2\pi r h)$.

The cost for the entire cylindrical box, then is: $C = C_1 + C_2$.

$$C = \pi r^2 + 5(2\pi rh)$$
$$= \pi r^2 + 10\pi rh$$

We can reduce this function to a function of one variable by using the fact that the volume of the cylinder is $25\,\text{ft}^3$.

$$V = 25 = \pi r^2 h \implies h = \frac{25}{\pi r^2}$$

$$C = \pi r^2 + 10\pi r \left(\frac{25}{\pi r^2}\right)$$
$$= \pi r^2 + \frac{250}{r} = \pi r^2 + 250 r^{-1}$$

The radius cannot be zero, because we would otherwise be dividing by zero, the domain is $(0, \infty)$.

$$C'(r) = 2\pi r - 250 r^{-2} = 0$$
$$2\pi r = \frac{250}{r^2}$$
$$2\pi r^3 = 250$$
$$r^3 = \frac{250}{2\pi} = \frac{125}{\pi}$$

$$\implies r = \sqrt[3]{\frac{125}{\pi}} = \frac{5}{\sqrt[3]{\pi}} \text{ or } 5\pi^{\frac{-1}{3}}$$

The dimensions should be

$$r = \frac{5}{\sqrt[3]{\pi}} \text{ ft}$$

and

$$h = \frac{25}{\pi r^2} = \frac{25}{\pi \cdot (5\pi^{\frac{-1}{3}})^2} = \frac{25}{\pi \cdot 25 \cdot \pi^{\frac{-2}{3}}}$$
$$= \frac{1}{\pi^{\frac{1}{3}}} = \frac{1}{\sqrt[3]{\pi}} \text{ ft}$$

Summary. The calculations for this problem are quite challenging. It is easier to keep fractional exponents until the very end, and then switch to radicals if you prefer. It is easier to calculate using fractional exponents.

Do not forget to answer the question, which always includes the units!

3.7. Rates of Change in Economics and the Sciences

Key Topics.
- Marginal Cost, Marginal Profit, Marginal Revenue
- Elasticity of Demand

Worked Examples.
- Using marginal cost
- Using derivatives for population studies
- Finding and interpreting the elasticity of demand

Overview

In this section we will use the derivative to measure changes in business, economics, and social sciences. The most important idea to understand here is that whenever we are talking about change in anything, we need to use the derivative. The derivative measures change in volume, cost, population, temperature, or any quantity which is changing.

Marginal Cost, Marginal Profit, Marginal Revenue. Another interpretation for the derivative used in business and economics is called marginals. The term *marginal cost* means the additional cost of producing one more unit. The definitions of *marginal revenue* and *marginal profit* parallel the *marginal cost* definition. We define these to be the **derivative** of the cost, profit, or revenue functions. In other words, $MC(X) = C'(x)$, $MP(x) = P'(x)$, and $MR(x) = R'(x)$.

Elasticity of Demand. Elasticity measures how **responsive** demand is to changes in price. If $E < -1$ we say that the demand is elastic which tells us that a small price cut will result in a relatively large increase in demand which in turn may result in a rise in revenue. On the other hand, if $E > -1$ we say that the demand is inelastic which tells us that a small price cut brings only a slight increase in demand which in turn can cause a decrease in revenue.

Example 1: Marginal Cost Problem

A company manufactures calculators and finds that its cost function, which is the total cost in dollars of manufacturing x calculators is

$$C(x) = 400 + 300\sqrt{x}$$

    a)   Find the marginal cost function.
    b)   Find the marginal cost when 81 calculators have been produced and interpret your answer.
    c)   Use the cost function to find out exactly how much the $81st$ calculator cost.

Strategy. Since we are asked for marginal cost, we know we need to find the derivative of the cost function. For the next part, we need to find $C'(81)$ which will tell us approximately how much more the $81st$ calculator costs than the total cost of 80 calculators. For the last part, we will calculate $C(81) - C(80)$ and see how close the derivative actually came to the exact answer. Because the derivative is a limit (remember?), it is an approximation, hopefully a very good approximation, but it **is** an approximation.

Solution.

$$\begin{aligned} C(x) &= 400 + 300\sqrt{x} = 400 + 300x^{\frac{1}{2}} \\ C'(x) &= 150x^{\frac{-1}{2}} = \frac{150}{\sqrt{x}} \\ C'(81) &= \frac{150}{\sqrt{81}} = \frac{150}{9} \approx \$16.67 \end{aligned}$$

This tells us that the approximate cost of the $81st$ calculator is $16.67.

Next, using the cost function and no calculus, let's find the exact cost of the $81st$ calculator.
The total cost for 81 calculators is $C(81) = 400 + 300\sqrt{81} = \$3100$.
The total cost for 80 calculators is $C(80) = 400 + 300\sqrt{80} \approx \$3083.28$.
Therefore, the cost of the $81st$ calculator is $\$3100 - \$3083.28 = \$16.72$
Using the marginal cost, $C'(81)$, the cost was only 5 cents less than the actual cost. Pretty close approximation, don't you agree?

Summary. This example shows how the approximation reached using the marginal cost is often quite close to the actual cost. The same conclusion holds true for marginal profit and marginal revenue.

Example 2: Population Study

A publisher who publishes children's books forecasts that the number of children in the United States x years from now will be:

$$P(x) = 10,000,000 - 10,000x + 500x^2 + 50x^3$$

Find the rate of change in the number of children
- a) x years from now
- b) 5 years from now and interpret your answer
- c) 6 years from now and interpret your answer

Strategy. Once again, because we are asked for the rate of change, we will find the derivative. We will then evaluate the derivative for $x = 5$ and $x = 6$. If the derivative is positive, we will conclude an increasing population of children. If the derivative is negative, then we will conclude a decreasing population of children.

Solution. a)
$$P(x) = 10,000,000 - 10,000x + 500x^2 + 50x^3$$
$$P'(x) = -10,000 + 1,000x + 150x^2$$

b)
$$P'(5) = -10,000 + 1,000(5) + 150(5^2)$$
$$= -1250$$

Conclusion: In 5 years the number of children in the US will be **decreasing** at the rate of 1250 children per year.

c)
$$P'(6) = -10,000 + 1,000(6) + 150(6^2)$$
$$= 1400$$

Conclusion: In 6 years the number of children in the US will be **increasing** at the rate of 1400 children per year.

Summary. This is just one more example of using the derivative to measure rates of change. By now, you are, hopefully, understanding that the derivative measures change!

Example 3: Elasticity of Demand

A commuter train estimates the demand function for its daily commuter tickets to be $D(p) = 100 - p^2$ (in thousands of tickets), where p is the price in dollars ($0 \leq p \leq 10$). Find the elasticity of demand when the price is a) \$4 and b) \$7.

Strategy. Using the formula for elasticity, $E(p) = \frac{pD'(p)}{D(p)}$, we will find the value for E which will tell us how elastic it is. Then we can interpret what happens at each of the two prices.

Solution.

$$\begin{aligned} E(p) &= \frac{pD'(p)}{D(p)} \\ &= \frac{p(-2p)}{100 - p^2} \\ &= \frac{-2p^2}{100 - p^2} \end{aligned}$$

a)
$$E(4) = \frac{-2(4^2)}{100 - 4^2} = \frac{-32}{84} = \frac{-8}{21}$$

Therefore, when $p = \$4$, the demand is inelastic, since $E > -1$

b)
$$E(7) = \frac{-2(7^2)}{100 - 7^2} = \frac{-98}{51} \approx -1.92$$

Therefore, when $p = \$7$, the demand is elastic, since $E < -1$.

Summary. For part a) when the price is \$4, an elasticity of $\frac{-8}{21}$ means that a 1% price change will only cause about an $\frac{8}{21}$% change in demand. Because it is such a small change, we say the demand is inelastic.

However, in part b), when the price is \$7, the elasticity is close to -2. This means that a 1% price change will cause almost a 2% change in demand. See the difference in the percentage of the change in demand?

3.8. Related Rates and Parametric Equations

Key Topics.
- Related Rates
- Parametric Equations

Worked Examples.
- A Ladder Problem
- A Shadow Problem
- A Related Rate Problem Involving an Angle
- Parametric Equations of an Ellipse

Overview

Related rates are simply rates which are related! Sounds strange, but that's what they are. Parametric equations will have x and y in them, but instead of y being a function of x, both are functions of a third variable.

Related Rates. By now you understand that the derivative measures the rate of change. The key word in the previous sentence is **rate**. As always, the first thing we need to do when we read a word problem is to translate the words into mathematical symbols. Whenever a quantity is changing, the words are translated into a derivative, often with respect to time. Words that tell of change are words similar to increase, decrease, getting larger/smaller, emptying, filling up etc. So, for example, if air is being pumped into a balloon at a rate of 3 cubic inches per second, we would write $\frac{dV}{dt} = 3$, since the volume is increasing. In these problems we will need a formula relating all the variables used in the problem. Next, we take the derivative with respect to time on both sides of the equation, using implicit differentiation. We will have numerical values for all the variables, except the rate that we are looking for. Then we substitute and finish the problem by solving the equation for the unknown quantity.

Parametric Equations. Often it is not enough to simply have y as a function of x. They may both be a function of another variable, often time, t. We need this because time often is an important factor in the problem. For example, we may need two quantities not only to arrive somewhere, but they must get there at the same time!

However, remember, if we are looking for a slope and we are using parametric equations, we still need $\frac{dy}{dx}$ since slope is always $\frac{\text{change in } y}{\text{change in } x}$. If you have $x(t)$ and $y(t)$, and you find $\frac{dx}{dt}$ and $\frac{dy}{dt}$, then we can find $\frac{dy}{dx}$ by using:

$$\frac{\frac{dy}{dt}}{\frac{dx}{dt}} = \frac{dy}{dt} \cdot \frac{dt}{dx} = \frac{dy}{dx}$$

Example 1: A Ladder Problem

A ladder 5 ft long is leaning against a vertical wall. If the base of the ladder is being pushed towards the wall at the rate of 1 foot per second, how fast will the top of the ladder move up the wall when the upper end of the ladder is 4 ft from the ground?

Strategy. When we sketch this picture of a ladder leaning against a building, we have a right triangle. We have information about the sides of this triangle. We know only **one** formula relating the sides of a right triangle - the Pythagorean Theorem. We know that $x^2 + y^2 = r^2$ if x and y are the legs of the triangle and r is the hypotenuse.

Solution. We start this problem by drawing a picture of the situation. Here we get a right triangle. Let's label all sides. It is important that the first picture is a "**moving**" picture, with arrows showing the direction of the movement.

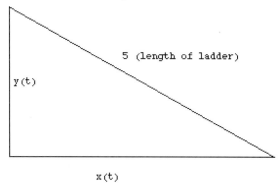

Moving picture

We need to organize our information. What information is **given** in this problem? The ladder is moving towards the wall at a rate of 1 ft/s. Note that $x(t)$ in this case is getting smaller, and hence the rate of change must be negative, that is

$$\frac{dx}{dt} = -1$$

We need to **find** $\frac{dy}{dt}$, the rate of change of $y(t)$ which is getting larger. We should, therefore, end up with a positive answer. The **instant** when $y = 4$ is the exact time we need. Let's draw a picture to show the relationship of all the variables at this instant. Using the Pythagorean theorem we find that $x = 3$.

3.8. RELATED RATES AND PARAMETRIC EQUATIONS

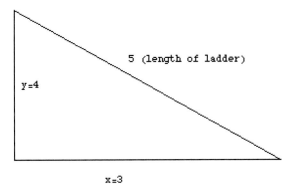

Instant picture

Relate the variables from the moving picture.

(Why not from the instant picture? Because the quantities in the instant picture are all fixed, and all rates of change are zero.)

$$x^2 + y^2 = 5^2 = 25$$

Now differentiate this equation with respect to time t.

$$2x\frac{dx}{dt} + 2y\frac{dy}{dt} = 0$$

Now (and not earlier!) we substitute in the numbers that are given, including the numbers from the instant when $y = 4$. The only unknown left will be $\frac{dy}{dt}$ so we will then be able to solve for it.

$$2 \cdot 3 \cdot (-1) + 2 \cdot 4 \cdot \frac{dy}{dt} = 0$$
$$-6 + 8\frac{dy}{dt} = 0$$
$$\frac{dy}{dt} = \frac{6}{8} = \frac{3}{4} \text{ ft/s}$$

Note that the rate at which the ladder is moving up the wall is positive, because the length y is increasing.

Summary. The answer must include units. It is very important to answer the question which is asked.

As you can see, you really need to stay organized. Sketch a picture of the changing (or moving) quantities. Label the parts, in this case the sides of a right triangle. **Write down** the translation of the problem from English into mathematical symbols. Once you have done all of the above, differentiate, substitute, and solve.

Example 2: A Shadow Problem

Suppose that a bear 6 ft tall is walking along a straight path with a ground speed of 2 ft/s towards a search light that is at the top of a 15 ft pole. How fast is the tip of the bear's shadow moving along the ground when the bear is 25 feet from the light?

Strategy. This is definitely a more challenging problem than the previous example. The 15 ft pole is perpendicular to the ground and we are dealing with distances from the pole which tells us that we are going to be working with right triangles again. A picture is necessary.

Solution. Let $x(t)$ denote the distance of the bear from the pole at time t, and let $s(t)$ denote the length of the bear's shadow.

Note: $x + s$ gives the distance from the pole to the tip of the shadow.

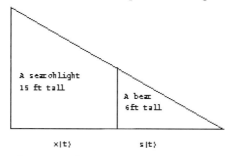

We have a picture of two similar triangles. What information is **given**? The bear is moving towards the searchlight, therefore his distance from the searchlight is decreasing making the derivative negative. In mathematical symbols, we write:

$$\frac{dx}{dt} = -2\,\text{ft/s}$$

We are asked to find $\frac{d}{dt}(x+s) = \frac{dx}{dt} + \frac{ds}{dt}$ at the instant when $x = 25$ ft (Why not $s'(t)$? Because $s'(t)$ measures the rate at which the shadow is decreasing).

We have $\frac{dx}{dt}$, so we need to find $\frac{ds}{dt}$ and then add them together. To relate the variables we can use ratios from the similar triangles, giving:

$$\frac{15}{x+s} = \frac{6}{s}$$
$$15s = 6x + 6s$$
$$9s = 6x$$
$$3s = 2x$$
$$s = \frac{2x}{3}$$

Differentiating with respect to time t, we get

$$\frac{ds}{dt} = \frac{2}{3}\frac{dx}{dt}$$

Now substituting (-2) for $\frac{dx}{dt}$, we get that
$$\frac{ds}{dt} = \frac{2}{3}(-2)$$
$$\frac{ds}{dt} = \frac{-4}{3} \text{ ft/s}$$
Thus
$$\frac{d}{dt}(x+s) = \frac{dx}{dt} + \frac{ds}{dt} = -2 + \frac{-4}{3} = -\frac{10}{3} \text{ ft/s}$$
and the tip of the bear's shadow is moving along the ground at $\frac{10}{3}$ ft/s.

Note that this problem did not require an *instant* picture.

Summary. The answer is negative because the tip of the shadow is moving closer to the searchlight and, hence, the distance is decreasing. It is necessary here, as in all word problems, to make sure that your answer makes sense.

Example 3: A Related Rate Problem Involving an Angle

An ostrich is running along a straight path at a speed of 4 ft/s. A searchlight is located on the ground 20 ft from the path. The searchlight is kept focused on the ostrich. At what rate is the searchlight rotating when the ostrich is 15 ft from the point on the path closest to the searchlight, assuming the ostrich is running in the direction **away** from the closest point?

Strategy. Once again, we start by sketching a picture of the problem. The sketch helps us see the placement of the variables and may give us a hint as to how they are related. In this problem, we again have a right triangle, but this time we are dealing with an angle, not just sides. A right triangle and an angle leads us to think of using the trigonometric definitions.

Solution. Once more, we start by drawing a "moving" picture. Denote by $x(t)$ the distance of the ostrich to the closest point on the straight path, and by $\theta(t)$ the angle of the searchlight with the line as drawn.

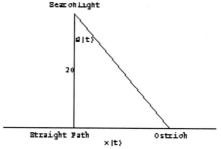

We are **given** that $\dfrac{dx}{dt} = 4\,\text{ft/s}$, a positive number since the ostrich is moving further away from the closest point and thus the distance is increasing and the derivative is positive. We need to find $\dfrac{d\theta}{dt}$ at the instant when $x = 15\,\text{ft}$. We can relate x and θ by using the tangent of θ, and

$$\tan(\theta) = \frac{\text{opposite side}}{\text{adjacent side}} = \frac{x}{20}$$

Differentiating this equation with respect to time, we get

$$\sec^2\theta \frac{d\theta}{dt} = \frac{1}{20}\frac{dx}{dt}$$

In this example we have to find the values of θ and $\sec^2\theta$ at the instant when $x = 15$. Let's draw a picture of this instant.

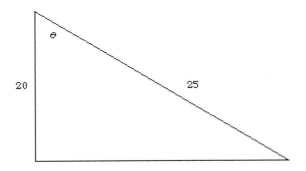

We can find $\sec^2\theta$ by using $\sec\theta = \dfrac{\text{hypotenuse}}{\text{adjacent}}$. Here $\sec\theta = \dfrac{25}{20} = \dfrac{5}{4}$. Now we can substitute the numbers into our above equation:

$$\left(\frac{5}{4}\right)^2 \frac{d\theta}{dt} = \frac{1}{20} \cdot 4 = \frac{1}{5}$$

$$\frac{d\theta}{dt} = \frac{1}{5} \cdot \left(\frac{4}{5}\right)^2 = \frac{16}{125} \text{ rad/s}$$

Hence the searchlight rotates at a rate of $\dfrac{16}{125}$ rad/s when the ostrich is 15 ft away from the closest point on the path to the searchlight.

Summary. Remember that most of the time when you are using trig functions, you are using radians, not degrees. If you are using a calculator, make sure that you are in radian mode. In this example, then, the units are radians per second.

Example 4: Parametric Equations of an Ellipse

A spider is repairing its spiderweb. The spider's position can be described by the following equations. The units for x and y are inches, and time t is measured in seconds.

$$\text{position} = \begin{cases} x(t) = 3\sin(\pi t) \\ y(t) = 2\cos(\pi t) \end{cases}$$

Find the spider's velocity and speed at time $t = 1\,\text{s}$, and describe the spider's motion.

Strategy. If you want to graph this on a graphing calculator, change the mode to parametric mode. If you then graph $x(t)$ and $y(t)$ you can see the ellipse.

We are given the position function. The change in position, the derivative of the position, always gives the velocity. Velocity has both magnitude (size) and direction. Speed, on the other hand, has only magnitude. For example, if you are driving 55 m.p.h. east, you have specified the velocity. If you only say you are driving 55 m.p.h., then you specified the speed. Speed can usually be found by taking the absolute value of the velocity.

Solution. We first find the velocity of the spider, and then the velocity and speed at time $t = 1$:

$$\text{velocity} = \begin{cases} \frac{dx}{dt} = 3\pi\cos(\pi t) \\ \frac{dy}{dt} = -2\pi\sin(\pi t) \end{cases}$$

$$\text{velocity at time } t = 1: \begin{cases} x'(1) = 3\pi\cos(\pi) = -3\pi \\ y'(1) = -2\pi\sin(\pi) = 0 \end{cases}$$

$$\text{speed} = \sqrt{[x'(1)]^2 + [y'(1)]^2} = \sqrt{[-3\pi]^2 + 0^2} = \sqrt{9\pi^2} = 3\pi\,\text{in/s}$$

The spider's path along the spider web

Summary. The starting point of the spider (time = 0) is:

$$x = 3\sin(\pi t) = 3\sin 0 = 0$$
$$y = 2\cos(\pi t) = 2\cos 0 = 2$$

Written as a point on the graph, we have $(0, 2)$, which tells us that it starts on the y-axis. We now know also, that at time $t = 1$ second, it is moving to the left at the rate of $3\pi\,\frac{\text{in}}{\text{s}}$.

CHAPTER 4

Integration

4.1. Area Under A Curve

Key Topics.
- Summation Notation
- Approximating Areas with Rectangles
- Riemann Sums
- Computing Areas Exactly

Worked Examples.
- Summation Examples
- Approximate Area under a Curve
- Computing the Area Exactly

Overview

One of the major problems over the centuries was how to find areas of shapes that were not circles, rectangles, trapezoids or any other shape that have simple formulas for the area. One method involved approximating the shape with lots of rectangles and then adding up the areas of all the rectangles.

Summation Notation. A capital sigma, Σ, is always used to denote summation. The index of summation is used to tell us where to begin and where to end. For example, $\sum_{i=1}^{5} 3i$ tells us to let $i = 1$, $i = 2$, $i = 3$, $i = 4$, and $i = 5$ in the formula $3i$ and then add up all our answers.

Approximating Areas with Rectangles. The area of a rectangle is easy to compute. Just multiply the length times the width. We are going to estimate the area of the region between a function and the x-axis by drawing lots of rectangles and adding all their areas together.

Riemann Sums. A mathematician in the 19*th* century, Bernhard Riemann, said that one way to estimate the area between a curve and the x-axis was to partition the interval $[a, b]$ (which the function covered) into $(x_0, x_1, ..., x_n)$. Each interval has width $\Delta x = x_i - x_{i-1} = \dfrac{b-a}{n}$. If we take any point c_i in the subinterval $[x_{i-1}, x_i]$, then the **Riemann sum of f** is given by $\sum_{i=1}^{n} f(c_i)\Delta x$.

If $f(x) \geq 0$ for all x, and if we think of the subinterval as the width of a rectangle, then $f(c_i)$ gives the length of the rectangle. Consequently, $f(c_i)\Delta x$ is simply the width times the length of a rectangle. If we do this for all i as i goes from 1 to n, we will be adding up the areas of all the rectangles and have an approximation of the area between the curve and the x-axis. It should be apparent that as n gets bigger and bigger, the approximation gets better and better.

Computing Areas Exactly. If we let $n \to \infty$ in the Riemann sum we will obtain the exact area $= \lim\limits_{n \to \infty} \sum_{i=1}^{n} f(x_i)\Delta x$.

Example 1: Summation Examples

a) Write out all terms and compute the sum

$$\sum_{i=1}^{4} \frac{24}{i}$$

b) Use summation rules to compute the sum

$$\sum_{i=1}^{30} \left(2i^2 - i + 4\right)$$

Strategy. In part a) substitute the numbers 1, 2, 3, and 4 for i in $\frac{24}{i}$ and add up all the numbers.

Part b) we can work the same way as part a) using all the integers from 1 to 30. However, it is much easier to use the summation rules:

$$\sum_{i=1}^{n} c = cn, \quad \sum_{i=1}^{n} i = \frac{n(n+1)}{2}, \quad \text{and} \quad \sum_{i=1}^{n} i^2 = \frac{n(n+1)(2n+1)}{6}$$

Solution. a)

$$\sum_{i=1}^{4} \frac{24}{i} = \frac{24}{1} + \frac{24}{2} + \frac{24}{3} + \frac{24}{4} = 24 + 12 + 8 + 6 = 50$$

b)

$$\sum_{i=1}^{30} i^2 = \frac{30(30+1)(2 \cdot 30 + 1)}{6} = 9,455$$

$$\sum_{i=1}^{30} i = \frac{30(30+1)}{2} = 465$$

$$\sum_{i=1}^{30} 4 = 30 \cdot 4 = 120$$

Thus we have,

$$\begin{aligned} \sum_{i=1}^{30} \left(2i^2 - i + 4\right) &= 2\sum_{i=1}^{30} i^2 - \sum_{i=1}^{30} i + \sum_{i=1}^{30} 4 \\ &= 9,455 + 465 + 120 \\ &= 10,040 \end{aligned}$$

Summary. The sigma notation is simply a shorthand way of telling us to substitute consecutive integers for the variable and add up all the answers.

Example 2: Approximate Area under a Curve

Approximate the area under the curve $f(x) = 2x^2 + 1$ on the interval $[0, 4]$. Use left-endpoint and right-endpoint approximations for $n = 2$ and $n = 4$, and use midpoint approximation for $n = 2$.

Strategy. If $n = 2$, we partition $[0, 4]$ into two equal intervals. The width of each interval is $\Delta x = \dfrac{4-0}{2} = 2$. For the left endpoint approximation, we draw a rectangle so that the left side of the rectangle hits the curve. For example, the first interval is $[0, 2]$. The rectangle will touch the curve at $f(0) = 1$. For the right endpoint, we draw a rectangle so that the right side of the rectangle hits the curve. In the first interval, it would touch at $f(2) = 9$. For a midpoint, the rectangle would touch the graph at $f(1) = 3$. In all cases this is the length (or height) of the rectangle. The width is shown above as 2. If $n = 4$ we do the exact same process using 4 intervals.

Solution. $n = 2$:

x	0	2	4
$f(x)$	1	9	33

Thus the left-endpoint and right-endpoint approximations are:

$$L_2 = f(0) \cdot 2 + f(2) \cdot 2 = 1 \cdot 2 + 9 \cdot 2 = 20$$
$$R_2 = f(2) \cdot 2 + f(4) \cdot 2 = 9 \cdot 2 + 33 \cdot 2 = 84$$

The midpoints are $x = 1$ and $x = 3$:

x	1	3
$f(x)$	3	19

The midpoint approximation:

$$M_2 = f(1) \cdot 2 + f(3) \cdot 2 = 3 \cdot 2 + 19 \cdot 2 = 44$$

$n = 4$:

x	0	1	2	3	4
$f(x)$	1	3	9	19	33

Thus the left- and right-endpoint approximations for $n = 4$ are:

$$L_4 = f(0) \cdot 1 + f(1) \cdot 1 + f(2) \cdot 1 + f(3) \cdot 1 = 1 + 3 + 9 + 19 = 32$$
$$R_4 = f(1) \cdot 1 + f(2) \cdot 1 + f(3) \cdot 1 + f(4) \cdot 1 = 3 + 9 + 19 + 33 = 64$$

Summary. Note that the approximations using the left-endpoint, the right-endpoint, and midpoint are not very close to each other. This is because of the extremely small number,n , of rectangles used. In the next example, we will see a much better approximation when we let $n \to \infty$.

Example 3: Computing the Area Exactly

Find the exact area under the curve $f(x) = 2x^2 + 1$ on the interval $[0, 4]$.

Strategy. We discuss this example in a more general way than example 2. Here we use n intervals and then let $n \to \infty$. It involves more abstract thinking than example 2 needed.

Solution. Using n subintervals, each of length $\Delta x = \dfrac{4-0}{n} = \dfrac{4}{n}$. $x_0 = 0$, $x_1 = \dfrac{4}{n}$, $x_2 = \dfrac{8}{n}$. Following this pattern $x_i = \dfrac{4i}{n}$. In this case

$$f(x_i) = 2\left(\frac{4i}{n}\right)^2 + 1 = \frac{32i^2}{n^2} + 1$$

$$f(x_i)\Delta x = \left(\frac{32i^2}{n^2} + 1\right)\left(\frac{4}{n}\right) = \frac{128i^2}{n^3} + \frac{4}{n}$$

Using the right-endpoint approximation for n subintervals, we get

$$\begin{aligned}
A_n &= \sum_{i=1}^{n} f(x_i)\Delta x \\
&= \sum_{i=1}^{n}\left(\frac{128i^2}{n^3} + \frac{4}{n}\right) \\
&= \sum_{i=1}^{n} \frac{128i^2}{n^3} + \sum_{i=1}^{n} \frac{4}{n}
\end{aligned}$$

(n is a constant here, rewrite summation)

$$= \frac{128}{n^3}\sum_{i=1}^{n} i^2 + \frac{1}{n}\sum_{i=1}^{n} 4$$

(use the summation rules)

$$= \frac{128}{n^3}\left(\frac{n(n+1)(2n+1)}{6}\right) + \frac{1}{n}(4n)$$

$$= \frac{64}{3}\left(\frac{n(n+1)(2n+1)}{n^3}\right) + 4$$

Finally, we can compute the limit of these approximations. We have

$$\begin{aligned}
\lim_{n\to\infty} A_n &= \lim_{n\to\infty}\left(\frac{64}{3}\left(\frac{n(n+1)(2n+1)}{n^3}\right) + 4\right) \\
&= \lim_{n\to\infty} \frac{64}{3}\left(\frac{2n^3 + 3n^2 + n}{n^3}\right) + 4 \\
&= \left(\frac{64}{3}\cdot 2 + 4\right) = \frac{140}{3}
\end{aligned}$$

Summary. In order to find the exact area, we had to partition the interval into n sub-intervals each of width $\Delta x = \dfrac{4-0}{n} = \dfrac{4}{n}$. Then we had to find the length of each rectangle using $f(x_i) = \dfrac{32i^2}{n^2} + 1$. Taking length times width we obtained the area of each rectangle as $\left(\dfrac{32i^2}{n^2} + 1\right)\left(\dfrac{4}{n}\right)$. We then took the sum of n such rectangles and let $n \to \infty$. We were able to manipulate the expressions so that we could utilize the summation rules. Remember if i is the index of summation, then any other letter (namely n in this problem) acts as a constant and can be brought in front of the summation sign.

4.2. The Definite Integral

Key Topics.
- The Definite Integral as Limit of Riemann Sums
- The Definite Integral as Signed Area
- Properties of Definite Integrals
- Overall Change and Distance Traveled
- Average Value of Functions
- Integral Mean Value Theorem

Worked Examples.
- Evaluating a Definite Integral using Riemann Sums
- Evaluating a Definite Integral as Signed Area
- Finding Total Area
- Finding Distance Traveled
- Using the Properties of Definite Integrals
- Finding an Average Value
- Approximate an Integral using the Integral Mean Value Theorem

Overview

In this section we will define the definite integral and show its uses in finding an area.

The Definite Integral as Limit of Riemann Sums. In example 3 in section 4.1, we found the exact area under a curve (actually between the curve and the x-axis, but it is often said *under the curve*) using the limit of Riemann sums. In this section we call this limit the *Definite Integral*.

The Definite Integral as Signed Area. When we calculate an integral, we get a positive number if the region lies above the x-axis, and a negative number if the region lies below the x-axis. Consequently, if the region is composed of two identical regions with one above the x-axis, and the other below the x-axis, the integral will equal zero. Clearly, the area is not equal to zero. We have what we call a signed area, which tells us that the two areas cancelled each other out. A negative answer tells us that most of the region lies below the x-axis.

Properties of Definite Integrals. Note the properties stated in your text. Hopefully, you will find them quite logical. They tell us that the integral of a sum or difference of two functions is the same as the sum or difference of the integrals of the two functions. Also, we see that we can put a constant in front of the integral sign and integrate the remaining function.

The last property tells us that if we are integrating over an interval $[a,b]$, then we can separate the integral into two integrals at any point c in (a,b). For example,

$$\int_1^3 f(x)dx = \int_1^2 f(x)dx + \int_2^3 f(x)dx$$

Overall Change and Distance Traveled. Compare this to the signed distance. The signed distance tells us how far the end is from the beginning, this tells how much distance was actually covered. For example, if you live 2 miles from work, but you go to the grocery, then to the shoe store, then to the cleaners, and then to work, the *signed area* would be 2 miles, because that is how far you are from home. However, the distance traveled would include all the miles you went running all your errands.

Average Value of Functions. A function takes on many values over an interval. The average value is exactly that, the average of all the values that the function takes on.

Integral Mean Value Theorem. This theorem tells us that at some point on the interval, the value of a continuous function is exactly equal to its average value. This theorem is used for proofs of some important theorems.

Example 1: Evaluating a Definite Integral using Riemann Sums

Evaluate the definite integral by finding the limit of Riemann Sums:

$$\int_0^5 (3x - 6)\, dx$$

Strategy. This is very similar to example 3 in section 4.1. Only this time, we are giving the limit of the summation a name. We are calling it the **definite integral**.

Solution. Using n subintervals, each of length $\Delta x = \dfrac{5}{n}$, with $x_0 = 0$, $x_1 = \dfrac{5}{n}$, $x_2 = \dfrac{10}{n}$, and then following this pattern $x_i = \dfrac{5i}{n}$. In this case

$$f(x_i) = 3\left(\frac{5i}{n}\right) - 6 = \frac{15i}{n} - 6$$

$$f(x_i)\Delta x = \left(\frac{15i}{n} - 6\right)\left(\frac{5}{n}\right) = \frac{75i}{n^2} - \frac{30}{n}$$

Using the right-endpoint approximation for n subintervals, we get

$$\begin{aligned}
A_n &= \sum_{i=1}^{n} f(x_i)\Delta x \\
&= \sum_{i=1}^{n}\left(\frac{75i}{n^2} - \frac{30}{n}\right) \\
&= \sum_{i=1}^{n} \frac{75i}{n^2} - \sum_{i=1}^{n} \frac{30}{n}
\end{aligned}$$

(n is a constant, move n in front of $\sum$)

$$\begin{aligned}
&= \frac{75}{n^2}\sum_{i=1}^{n} i - \frac{1}{n}\sum_{i=1}^{n} 30 \\
&= \frac{75}{n^2}\left(\frac{n(n+1)}{2}\right) - \frac{1}{n}(30n) \\
&= \frac{75}{2}\left(\frac{n(n+1)}{n^2}\right) - 30
\end{aligned}$$

Finally, we can compute the limit of these approximations. We have

$$\lim_{n\to\infty} A_n = \lim_{n\to\infty}\left(\frac{75}{2}\left(\frac{n(n+1)}{n^2}\right) - 30\right) = \left(\frac{75}{2}\cdot 1 - 30\right) = \frac{15}{2}$$

Summary. This problem was done exactly the same way as example 3 in section 4.1 of this book. There is a very detailed description of the process if you want to go back and read it. Once again we used i, the index of summation as the variable, having n act as a constant. We can always bring constants in front of the summation sign and then use the summation rules.

Example 2: Evaluating a Definite Integral as Signed Area

Evaluate the integral $\int_0^5 (3x - 6)\, dx$ by finding the signed area under the graph $f(x) = 3x - 6$, $0 \leq x \leq 6$.

Strategy. In this problem, we find the area of each of the two triangles formed. The area of a triangle is $\frac{1}{2}$·base·height. The **signed** area is different from the total area. For signed area, the region below the x-axis is subtracted from the area of the triangle above the x-axis. It is quite possible to get a negative answer or zero when working with signed area.

Solution.

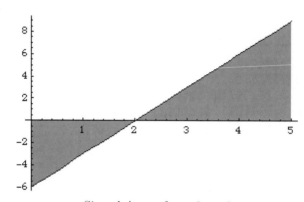

Signed Area of $y = 3x - 6$

$$\int_0^5 (3x - 6)\, dx$$
$$= \text{(area of triangle above the } x\text{-axis)} - \text{(area of triangle below the } x\text{-axis)}$$
$$= \frac{1}{2} \cdot 3 \cdot 9 - \frac{1}{2} \cdot 2 \cdot 6 = \frac{27}{2} - 6 = \frac{15}{2}$$

Summary. In $\int_0^5 (3x-6)dx$, the area of the region below the x-axis is subtracted from the area of the region above the x-axis. Consequently, the answer is less than the area of the region above the x-axis.

Example 3: Finding Total Area

Write the total area of the graph below as an integral, or as a sum of integrals.

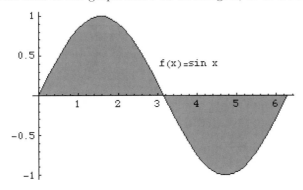

Strategy. Since we are looking for total area in this problem, we must subtract any *negative* area, thereby adding that area. Subtracting a negative number is the same as adding a positive number. For example, if we wanted to plant grass over the entire region, we would need all the areas to appear as positive numbers.

Solution. First note that $\int_0^{2\pi} \sin x \, dx = 0$, since the area above and below the x-axis are the same. To find the total area we must **subtract** the signed area on the interval from π to 2π since the signed area is negative. Note then that the area above the x-axis is given by the integral $\int_0^{\pi} \sin x \, dx$, and the area below by $-\int_{\pi}^{2\pi} \sin x \, dx$. The minus sign in the second integral is necessary since the integral itself is negative. Thus the total area, which is different from the signed area, is equal to:

$$\text{Total Area} = \int_0^{\pi} \sin x \, dx - \int_{\pi}^{2\pi} \sin x \, dx$$

Summary. When we are looking for total area, we must make sure that any part of the region which lies below the x-axis has a negative sign in front of the integral. Otherwise, the negative answer resulting from that integral will take away from the total area instead of adding to it.

Example 4: Finding Distance Traveled

Suppose the velocity of a train slowly traveling on straight tracks is given by $v(t) = 192 - 64t$ ft/s. Find the distance from the starting point at time $t = 0$, to the point 5 seconds later. In addition to that, find the total distance traveled.

Strategy. For the first part, the signed area, we will find the area of the upper triangle and the area of the lower triangle. We will then subtract the area of the lower from the area of the upper.

For the second part, the total distance, we will find the integral of the function over the interval $[0, 3]$, which is positive, and then subtract the integral of the function over the interval $[3, 5]$, which is negative. Note that we will be subtracting a negative number here and hence, will be adding the two areas.

Solution.

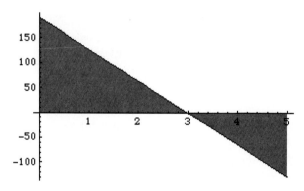

Velocity of the train

Note that the integral $\int_0^5 v(t)dt$ is equal to the distance from the starting point. Thus

$$
\begin{aligned}
\text{Distance} &= \int_0^5 v(t)dt \\
&= (\text{area of triangle above the } t\text{-axis}) - (\text{area of triangle below the } t\text{-axis}) \\
&= \frac{1}{2} \cdot 3 \cdot 192 - \frac{1}{2} \cdot 2 \cdot 128 \\
&= 288 - 128 = 160 \,\text{ft}
\end{aligned}
$$

The total distance traveled is equal to the sum of the areas of the triangles, or in terms of integrals $\int_0^3 v(t)dt - \int_3^5 v(t)dt$. Thus

$$
\begin{aligned}
\text{Total Distance} &= \frac{1}{2} \cdot 3 \cdot 192 + \frac{1}{2} \cdot 2 \cdot 128 \\
&= 416 \,\text{ft}
\end{aligned}
$$

Summary. Make sure you understand the two different concepts here. The first tells us the actual change from the starting position to the ending position. It doesn't tell us what route it took to get there. The total distance tells us the actual distance traveled to get to the endpoint.

Example 5: Using the Properties of Definite Integrals

If $\int_1^3 f(x)dx = 4$, $\int_1^5 f(x)dx = 6$, and $\int_5^3 g(x)dx = -2$, find $\int_3^5 (2f(x) - 3g(x))\, dx$.

Strategy. First we need to find $\int_3^5 f(x)dx$, which is equal to

$$\int_1^5 f(x)dx - \int_1^3 f(x)dx = 6 - 4 = 2$$

We also know that $\int_5^3 g(x)dx = -2$. Therefore, $-\int_3^5 g(x)dx = 2$. Now we have all the parts needed for the solution.

Solution.

$$\begin{aligned}
\int_3^5 (2f(x) - 3g(x))\, dx &= 2\int_3^5 f(x)dx - 3\int_3^5 g(x)dx \\
&= 2(2) - 3 \cdot (2) \\
&= 2 \cdot 2 - 6 \\
&= -2
\end{aligned}$$

Summary. In this problem we made use of the fact that

$$\int_b^a g(x)dx = -\int_a^b g(x)dx$$

In addition, make sure that you understand how

$$\int_1^5 f(x)dx - \int_1^3 f(x)dx = \int_3^5 f(x)dx$$

Example 6: Finding an Average Value

Find the average value of the function $f(x) = 3x - 6$ on the interval $[0, 5]$.

Strategy. Whenever we find an average of anything, we add up all the items and then divide by the number of items. Since the integral sign is an elongated S for sum, we find the sum by integrating. Then to obtain the average, we divide by the length of the interval $[a, b]$.

Solution. We already found that the integral $\int_0^5 (3x - 6) \, dx = \frac{15}{2}$ in examples 1 and 2. Thus,

$$\text{average value} = \frac{1}{5-0} \int_0^5 (3x - 6) \, dx = \frac{1}{5} \cdot \frac{15}{2} = \frac{3}{2}$$

Summary. As a student, you are used to getting your grades averaged. The teacher adds up all your scores and then divides by the total number of tests to find out your average. Finding the average value of a function over an interval is the same concept.

Example 7: Approximate an Integral using the Integral Mean Value Theorem

Use the Integral Mean Value Theorem to estimate the integral
$$\int_{-1}^{1} \frac{2dx}{x^2+1}$$

Strategy. The Integral Mean Value Theorem basically tells us that a continuous function will take on its average value at some point. It is a theorem of existence and will later be used in proofs of major theorems. We can also use this theorem to estimate the value of an integral.

Solution. Note that $x^2 + 1 \geq 1$, and $x^2 + 1 \leq 2$ for $-1 \leq x \leq 1$. Thus
$$1 \leq x^2 + 1 \leq 2$$
$$\implies 1 \geq \frac{1}{x^2+1} \geq \frac{1}{2}$$
$$\implies 2 \geq \frac{2}{x^2+1} \geq 1$$

That is $1 \leq f(x) \leq 2$ for $-1 \leq x \leq 1$. Since f is a continuous function on $[-1, 1]$, the Mean Value Theorem guarantees a number c between -1 and 1, with

$$f(c) = \frac{1}{b-a}\int_a^b f(x)dx = \frac{1}{1-(-1)}\int_{-1}^{1}\frac{2dx}{x^2+1} = \frac{1}{2}\int_{-1}^{1}\frac{2dx}{x^2+1}$$

Since f is bounded by 1 and 2, we know that $f(c)$ is between 1 and 2. So we have:
$$1 \leq f(c) \leq 2$$
$$1 \leq \frac{1}{2}\int_{-1}^{1}\frac{2dx}{x^2+1} \leq 2$$
$$2 \leq \int_{-1}^{1}\frac{2dx}{x^2+1} \leq 4$$

Thus the integral is some number between 2 and 4.

Summary. We have been approximating the values of integrals throughout the last two sections. In section 4.3, we will begin to actually find the values of the integrals without having to use the approximating sums.

4.3. Antiderivatives

Key Topics.
- Antiderivatives
- Power Rule
- Position-Velocity-Acceleration

Worked Examples.
- Using the Power Rule
- Functions of the Form $f(ax)$
- Functions of the Form $\frac{f'(x)}{f(x)}$
- Finding the Position of a Falling Object

Overview

In this section we learn how to find the function from the derivative of the function. We will see that we can find the area between a curve and the x-axis even when the shape is not a simple rectangle, circle, or any other shape with a simple area formula.

Antiderivatives. Just like we undo addition with subtraction, and we undo multiplication with division, we can undo differentiation. The process is called antidifferentiation. We will see that another name, and the one more commonly used, is integration with the result being called the integral.

Power Rule. Just as in differentiation, the power rule is used very often. Since we have to undo a derivative, we must use the opposite processes of differentiation. Because $\frac{d}{dx}(x^n) = nx^{n-1}$ which tells us to subtract the exponent by one, and to multiply the expression by n, when we integrate we need to reverse the procedure by adding one to the exponent, and dividing by the new exponent. Consequently,

$$\int x^n dx = \frac{x^{n+1}}{n+1} + c$$

The c is there because in $\int x^n dx$ we are looking for every possible function $f(x)$ such that $f'(x) = x^n$. For example, what if we had $\int 5x^4 dx$? We are asked to find every possible function with derivative $5x^4$. Look at $f(x) = x^5 + 7$. Then $f'(x) = 5x^4$, right? But what about $f(x) = x^5 - 38$? Then $f'(x) = 5x^4$, also. Since the derivative of a constant is zero, the derivative of $f(x) = x^5 + c$ for any constant c gives the derivation $5x^4$. Hence, we call this an indefinite integral, because we do not have information to find a particular constant. Every indefinite integral **must** have $+c$ in the answer.

Position-Velocity-Acceleration. We know that acceleration due to gravity is $-32\,\text{ft}/\text{s}^2$. Since acceleration is the derivative of the velocity, we can integrate acceleration to get the velocity function. Similarly, since velocity is the derivative of the position, we can integrate the velocity function to obtain the position function. We just go backwards from what we did in the derivative chapter.

Example 1: Using the Power Rule

Evaluate the following integrals.

a) $\int \left(2x^2 - 3x^6\right) dx$

b) $\int 3\sqrt{x}\, dx$

c) $\int \dfrac{3}{2\sqrt{x}}\, dx$

d) $\int \dfrac{2x^2 - 5x + x^{\frac{3}{4}}}{x^2}\, dx$

Strategy. The only rule we know at this time is the power rule. Therefore, each integral **must** be written in the form of the power rule, which is a single letter base to a numerical power.

Solution. a)

$$\int \left(2x^2 - 3x^6\right) dx = 2\left(\frac{1}{3}x^3\right) - 3\left(\frac{1}{7}x^7\right) + c$$
$$= \frac{2}{3}x^3 - \frac{3}{7}x^7 + c$$

b)

$$\int 3\sqrt{x}\, dx = \int 3x^{\frac{1}{2}}\, dx = 3\left(\frac{2}{3}x^{\frac{3}{2}}\right) + c$$
$$= 2x^{\frac{3}{2}} + c = 2\sqrt{x^3} + c$$

c)

$$\int \frac{3}{2\sqrt{x}}\, dx = \int \frac{3}{2}x^{-\frac{1}{2}}\, dx$$
$$= \frac{3}{2}\left(2x^{\frac{1}{2}}\right) + c$$
$$= 3x^{\frac{1}{2}} + c = 3\sqrt{x} + c$$

d)

$$\int \frac{2x^2 - 5x + x^{\frac{3}{4}}}{x^2}\, dx = \int \left(\frac{2x^2}{x^2} - \frac{5x}{x^2} + \frac{x^{\frac{3}{4}}}{x^2}\right) dx$$
$$\int \left(2 - \frac{5}{x} + x^{\frac{3}{4} - 2}\right) dx$$

$$= \int \left(2 - \frac{5}{x} + x^{-\frac{5}{4}}\right) dx$$
$$= 2x - 5\ln|x| - 4x^{-\frac{1}{4}} + c$$

Summary. Make sure you are completely comfortable working with exponents and converting to negative and fractional exponents. That process will be used repeatedly throughout the study of calculus. All formulas are given as powers.

4.3. ANTIDERIVATIVES

Example 2: Functions of the Form $f(ax)$

Evaluate

a) $\int \cos(5x)\, dx$

b) $\int \frac{\sin(2x)\, dx}{2}$

c) $\int e^{2x}\, dx$

d) $\int \frac{6\, dx}{1 + 4x^2}$

Strategy. We have to keep in mind that integration is the opposite of differentiation. Therefore, to check the answer to an integral, we differentiate our answer and expect to end up with the integrand.

Solution. When we integrate $\cos(5x)$, we need to think about what derivative is equal to $\cos(5x)$. The derivative of $\sin x$ is $\cos x$. In the integral the x is multiplied by 5. It is reasonable, then, to expect the integral to be multiplied by $\frac{1}{5}$. We can try it and see if it works!

a)
$$\int \cos(5x)\, dx = \frac{1}{5}\sin(5x) + c$$

We can check this by taking the derivative:
$$\frac{d}{dx}\left[\frac{1}{5}\sin(5x)\right] = \frac{1}{5} \cdot 5\cos(5x) = \cos(5x)$$

See how you need the $\frac{1}{5}$ there when you check your answer by taking the derivative?

b) Here, we have a similar example, thinking that we need a $\frac{1}{2}$ in the answer to the integral because of the $2x$ in the integrand.

$$\int \frac{\sin(2x)}{2}\, dx = \frac{1}{2}\int \sin(2x)\, dx$$
$$= \frac{1}{2} \cdot \frac{-1}{2}\cos(2x) + c = \frac{-1}{4}\cos(2x) + c$$

Check: $\frac{d}{dx}\left[\frac{-1}{4}\cos(2x)\right] = +\frac{1}{4} \cdot 2\sin(2x) = \frac{1}{2}\sin(2x)$

c) This integral has a $\frac{1}{2}$ in the answer, also.

$$\int e^{2x}\, dx = \frac{1}{2}e^{2x} + c$$

Check: $\dfrac{d}{dx}\left(\dfrac{1}{2}e^{2x}\right) = \dfrac{1}{2}e^{2x} \cdot 2 = e^{2x}$

d) Here, we have to first decide which type of antiderivative will work. The $1+4x^2$ in the denominator can be written as $1 + (2x)^2$. Then the integrand resembles the derivative of the arctan function. We probably need a $\dfrac{1}{2}$ again in the answer because of the $2x$. Let's try it and see.

$$\begin{aligned}
\int \dfrac{6}{1+4x^2}\,dx &= 6\int \dfrac{1}{1+4x^2}\,dx \\
&= 6 \cdot \dfrac{1}{2} \tan^{-1}(2x) + c \\
&= 3\tan^{-1}(2x) + c
\end{aligned}$$

Check: $\dfrac{d}{dx}(3\tan^{-1}(2x)) = 3 \cdot \dfrac{1}{1+(2x)^2} \cdot 2 = 6\left(\dfrac{1}{1+4x^2}\right) = \dfrac{6}{1+4x^2}$

Summary. Note how the derivatives of the answers give the integrands. In a later section in this chapter we will learn u-substitution which is a more straightforward method of working these problems.

Example 3: Functions of the Form $\frac{f'(x)}{f(x)}$

Evaluate

a) $\int \frac{\cos x \, dx}{\sin x}$

b) $\int \frac{2x+3}{x^2+3x+4} dx$

Strategy. In both of these problems the numerator is the derivative of the denominator. Think about how we differentiate the natural logarithm function.

Solution. a)
$$\int \frac{\cos x}{\sin x} dx = \int \frac{1}{\sin x} \cdot (\cos x) dx$$

Since $\frac{d}{dx}(\sin x) = \cos x$, we have 1 over a function times the derivative of that exact function which makes it the same form as when we took the derivative of the natural logarithm. Let's try the answer, and see if it works!

$$\int \frac{1}{\sin x} \cdot (\cos x) dx = \ln|\sin x| + c$$

Check: $\frac{d}{dx}(\ln|\sin x| + c) = \frac{1}{\sin x} \cos x = \frac{\cos x}{\sin x}$ ✓

b)
$$\int \frac{2x+3}{x^2+3x+4} dx = \int \frac{1}{x^2+3x+4} \cdot (2x+3) dx$$

Since $\frac{d}{dx}(x^2+3x+4) = 2x+3$, once again we have 1 over a function times the derivative of that function. Once again, then, let's try the natural logarithm and see if it works.

$$\int \frac{1}{x^2+3x+4} \cdot (2x+3) dx = \ln|x^2+3x+4| + c$$

Check: $\frac{d}{dx}\left(\ln|x^2+3x+4| + c\right) = \frac{1}{x^2+3x+4}(2x+3) = \frac{2x+3}{x^2+3x+4}$ ✓

Summary. These problems give us an understanding of the relationship between integration and differentiation. Once again, when we learn u-substitutions, the mechanics of these problems will be more straightforward.

Example 4: Finding the Position of a Falling Object

A chocolate pie is thrown directly upward at a speed of 64 ft/s from a platform 80 ft above the ground. At what time does the pie reach its highest point? How high does it go? When does it hit the ground? What is the velocity at that instant?

Strategy. It is important to remember here that velocity is the derivative of position, and that acceleration is the derivative of velocity. We can integrate to go back the other direction. Also, acceleration due to gravity is given by $a = -32\,\text{ft}/\text{s}^2$.

Solution. At what time does the pie reach the highest point? This asks for the maximum of the position function. We need to know the position function. We work backwards from $a = -32\,\text{ft}/\text{s}^2$ to the velocity function, and then integrate again to obtain the position function.

Further, we know that the initial velocity is 64 ft/s, that is $v(0) = 64$. In addition the initial height of the height is given by $s(0) = 80$.

$$v = \int a(t)\,dt = \int -32\,dt = -32t + c$$

Since $v(0) = 64$, we can find the value of the constant c:
$v(0) = -32 \cdot 0 + c = 64 \implies c = 64$.
Therefore
$$v = -32t + 64$$
From here we can find the height $s(t)$ of the pie:

$$s = \int v(t)\,dt = \int (-32t + 64)\,dt = -16t^2 + 64t + d$$

Since $s(0) = 80$, we can find the value of the constant d:
$s(0) = -16 \cdot 0 + 64 \cdot 0 + d = 80 \implies d = 80$
Thus

$$s = -16t^2 + 64t + 80$$

Now we can answer the questions:

a) At what time does the pie reach its highest point? It reaches its highest point when the velocity is zero, that is when $v = -32t + 64 = 0$, thus at time $t = 2\,\text{s}$.

The height after 2 seconds is $s(2) = -16 \cdot 2^2 + 64 \cdot 2 + 80 = 144\,\text{ft}$.

b) When does the pie hit the ground? It hits the ground when the height is zero, i.e.

$$\begin{aligned} s = -16t^2 + 64t + 80 &= 0 \\ -16\left(t^2 - 4t - 5\right) &= 0 \\ -16\left(t - 5\right)\left(t + 1\right) &= 0 \end{aligned}$$

The height is zero when $t = -1$ or $t = 5$. Therefore the pie reaches the ground after 5 seconds.

c) The velocity when it hits the ground is given by $v(5) = -32 \cdot 5 + 64 = -96\,\text{ft}/\text{s}$.

Summary. Knowing the acceleration due to gravity and the initial conditions gives us enough information to find the position and velocity functions. Then we can use those functions to answer the questions.

4.4. The Fundamental Theorem of Calculus

Key Topics.
- Fundamental Theorem of Calculus, Part I
- Fundamental Theorem of Calculus, Part II
- Total Change

Worked Examples.
- Evaluating Definite Integrals
- Using the Fundamental Theorem of Calculus, Part II
- Total Change in Some Population

Overview

With the Fundamental Theorem of Calculus we have a major improvement over computing integrals as a limit of Riemann sums, which many times couldn't be done even if we wanted to do it. The Fundamental Theorem of Calculus, Parts I and II, connect the two major concepts in calculus - the derivative and the integral.

Fundamental Theorem of Calculus, Part I. This theorem tells us how to evaluate a definite integral. Find an antiderivative $F(x)$. If b is the upper limit of the integral, and a is the lower limit, then we evaluate $F(b) - F(a)$. Symbolically:

$$\int_a^b f(x) = F(b) - F(a) \quad \text{where} \quad F'(x) = f(x)$$

Fundamental Theorem of Calculus, Part II. If the upper limit of the integral is a variable, and the lower limit is a constant, then the derivative of the integral with respect to the upper limit is simply the given integrand evaluated at its upper limit. At the lower limit, there was a constant so that when we took the derivative of it, it became zero.

The second part of the Fundamental Theorem is really one of the most important theorems in all of calculus. We use Part I more often in calculations, but Part II is the underlying concept. It says that differentiation and integration are inverse operations!

Think about it - when we evaluate a definite integral using an antiderivative, in many cases we have also geometrically found an area. The connection is quite incredible!

Total Change. In this part of the section we see that *total change* of a function over an interval is found by using the definite integral over the interval.

Example 1: Evaluating Definite Integrals

Evaluate

a) $\displaystyle\int_0^1 \left(4x^3 + e^x + 1\right) dx$

b) $\displaystyle\int_0^{16} \left(\frac{7}{4}x^{\frac{3}{4}} + 2\sqrt{x}\right) dx$

Strategy. We start by integrating (finding an antiderivative) the same way as we did for indefinite integrals. The only difference is that we do not need a $+c$ for a definite integral. Then following the Fundamental Theorem of Calculus, we evaluate the antiderivative, F, at the upper limit, b, and the lower limit, a, and subtract $F(b) - F(a)$.

Solution. a)

$$\int_0^1 \left(4x^3 + 2e^x + 1\right) dx = x^4 + 2e^x + x \Big|_0^1 = (1 + 2e + 1) - (0 + 2 + 0) = 2e$$

b)

$$\begin{aligned}
\int_0^{16} \left(\frac{7}{4}x^{\frac{3}{4}} + 2\sqrt{x}\right) dx &= \frac{7}{4}\left(\frac{4}{7}x^{\frac{7}{4}}\right) + 2\left(\frac{2}{3}x^{\frac{3}{2}}\right)\Big|_0^{16} \\
&= x^{\frac{7}{4}} + \frac{4}{3}x^{\frac{3}{2}}\Big|_0^{16} \\
&= (16)^{\frac{7}{4}} + \frac{4}{3}(16)^{\frac{3}{2}} - (0 + 0) \\
&= \left(\sqrt[4]{16}\right)^7 + \frac{4}{3}\left(\sqrt{16}\right)^3 \\
&= 2^7 + \frac{4}{3} \cdot 4^3 \\
&= 128 + \frac{256}{3} \\
&= \frac{384}{3} + \frac{256}{3} = \frac{640}{3}
\end{aligned}$$

Summary. As shown in these examples, if we can find an antiderivative of a function, then all we have to do is evaluate it at the upper and lower limits of integration. Be very careful - it is always evaluated top minus bottom.

Example 2: Using the Fundamental Theorem of Calculus, Part II

Find the following derivatives.

a) $\dfrac{d}{dx}\left(\displaystyle\int_\pi^x \sin^4 t\, dt\right)$

b) $\dfrac{d}{dx}\left(\displaystyle\int_\pi^{e^x} \sin^4 t\, dt\right)$

c) $\dfrac{d}{dx}\left(\displaystyle\int_{x^2+1}^\pi \sin^4 t\, dt\right)$

Strategy. If we could actually work out the problem showing every step, we would first find an antiderivative. Then we would evaluate the antiderivative at the upper and lower limits, and finally we would differentiate the antiderivative. This means that when the bottom limit is constant, the derivative is zero. All that is left is to differentiate the antiderivative evaluated at a function of x, giving us the same function, which we started with, as the integrand except it has been evaluated at some function of x, say $g(x)$. So in all of the above, we are going to end up with $\sin^4(g(x)) \cdot g'(x)$.

Solution. a)
$$\frac{d}{dx}\int_\pi^x \sin^4 t\, dt = \sin^4 x$$

b) Here we use the chain rule, $g(x) = e^x$:
$$\frac{d}{dx}\int_\pi^{e^x} \sin^4 t\, dt = \sin^4\left(e^x\right) \cdot \frac{d}{dx}\left(e^x\right) = \sin^4\left(e^x\right) e^x$$

c) The first step here is to reverse the order of the limits, so that the variable is on top. Then we use the chain rule for $g(x) = x^2 + 1$.

$$\begin{aligned}
\frac{d}{dx}\int_{x^2+1}^\pi \sin^4 t\, dt &= -\frac{d}{dx}\int_\pi^{x^2+1} \sin^4 t\, dt \\
&= -\sin^4\left(x^2+1\right) \cdot \frac{d}{dx}\left(x^2+1\right) \\
&= \sin^4\left(x^2+1\right)(-2x) \\
&= -2x\sin^4\left(x^2+1\right)
\end{aligned}$$

Summary. The mechanics of working these problems are not terribly difficult. However, try to appreciate the fact that it took hundreds of years before the connection between the integral, derivatives, antiderivatives and area were figured out. This didn't occur until the 17th century. It was a major advance in science and mathematics and gave us tools that were and still are used in the understanding of our universe.

Example 3: Total Change in Some Population

A squirrel population is increasing at a rate of $20+6t$ per year, where t is measured in years. By how much does the animal population increase between the fifth and tenth year?

Strategy. Since the problem has information about a **rate** that is increasing, we have a rate of change, or a derivative. Therefore, to get back to the original function, we need to *undo* the derivative, which means we must integrate over the interval from 5 to 10 to find how much the squirrel population increases over the period between the 5th year and the 10th year.

Solution.

$$\begin{aligned}
\text{Total Population Increase} &= \int_5^{10} (20+6t)\, dt \\
&= 20t + 3t^2 \Big|_5^{10} \\
&= (200 + 300) - (100 + 75) \\
&= 500 - 175 = 325 \text{ squirrels}
\end{aligned}$$

Summary. This problem is just one example of using the definite integral to measure the total change over a period of time. In the text there are other similar applications where the definite integral is used to find the total change over a given interval.

4.5. Integration by Substitution

Key Topics.
- Integration by Substitution
- Substitution in a Definite Integral

Worked Examples.
- A Substitution Involving a Trigonometric Function
- A Substitution Involving a Square Root
- A Substitution with a Square Root in the Denominator
- Where the Numerator is the Derivative of the Denominator
- A Definite Integral with a Logarithm
- A Definite Integral with an Inverse Trigonometric Function
- A Substitution that leads to an Expansion of the Integrand

Overview

So far we have integrated certain functions, but not most functions. As with differentiation, we started out with *simple* functions and then we moved to more complicated ones. We know, for example, how to integrate $\int x^{10} dx$, but we do not have a formula for $\int 4x^2(2x^3 + 7)^{10} dx$. We could, theoretically, multiply it all out, but none of us would like to do that! In this section we introduce integration by substitution, which will allow us to integrate many more functions, just as the chain rule allowed as to differentiate many more functions.

Integration by Substitution. What we must remember here is that in order to use any of the integration formulas, the integrand must be in **exactly** the same form as the integrand is in the formula. For example, we have $\int \sin x \, dx = -\cos x + c$, which has the sine of the variable x. It is **not** $\int \sin(4x) \, dx$ or $\int \sin(x^2) \, dx$. In these integrals, we first need to use a substitution to transform the integrand so that it looks the way we need it **before** we integrate.

When we substitute u for an expression containing x, then we also need to differentiate each side so that we know what to substitute for dx. You **cannot** integrate if the integrand contains both x and u.

Substitution in a Definite Integral. When we have an integral $\int_a^b f(x)dx$, then a and b are the upper and lower limits of the variable, x. The integrand contains "dx" (with respect to x) which tells us that the limits are x-limits. If we change the variable of integration to u, then we must change the limits to u-limits. So, for example, if we had $\int_1^2 f(x)dx$ and we made a substitution like $u = x + 4$, then the new integral would look like $\int_5^6 g(u)du$. The new limits came from saying, if $x = 1$, then $u = 1 + 4 = 5$, and if $x = 2$, then $u = 2 + 4 = 6$.

Example 1: A Substitution Involving a Trigonometric Function

Use a substitution to integrate

$$\int x^2 \sin\left(x^3\right) dx$$

Strategy. We know how to integrate $\sin u$ (since the angle is represented by u, a single letter). Therefore

$$\begin{aligned} \text{let } u &= x^3 \\ \text{then } du &= 3x^2 dx \\ \text{or } \tfrac{1}{3} du &= x^2 dx \end{aligned}$$

Solution.

$$\begin{aligned} \int x^2 \sin\left(x^3\right) dx &= \int \underbrace{\sin\left(x^3\right)}_{\sin u} \underbrace{x^2 dx}_{\tfrac{1}{3} du} \\ &= \int (\sin u) \cdot \tfrac{1}{3} du \\ &= \tfrac{1}{3} \int \sin u \, du \\ &= -\tfrac{1}{3} \cos u + c \\ &= -\tfrac{1}{3} \cos\left(x^3\right) + c \end{aligned}$$

Summary. Do not forget to substitute the x-variable back into the answer. The u-substitution was used to help us integrate, but the variable in the answer must match the variable in the given problem.

Example 2: A Substitution Involving a Square Root

Use a substitution to integrate

$$\int 5x^2 \sqrt{3x^3 + 5}\, dx$$

Strategy. First we can rewrite the integral with the 5 in front of the integral and with the radical sign replaced with a fractional exponent.

$$\int 5x^2 \sqrt{3x^3 + 5}\, dx = 5 \int x^2 (3x^3 + 5)^{\frac{1}{2}}\, dx$$

Solution.

$$\begin{aligned}
\text{Let } u &= 3x^3 + 5 \\
\text{then } du &= 9x^2\, dx \\
\text{or } \tfrac{1}{9} du &= x^2\, dx
\end{aligned}$$

$$\begin{aligned}
\int 5x^2 \sqrt{3x^3 + 5}\, dx &= 5 \int \underbrace{(3x^3 + 5)^{\frac{1}{2}}}_{u^{\frac{1}{2}}} \underbrace{x^2\, dx}_{\frac{1}{9} du} \\
&= 5 \int u^{\frac{1}{2}} \cdot \tfrac{1}{9} du \\
&= \tfrac{5}{9} \int u^{\frac{1}{2}}\, du \\
&= \tfrac{5}{9} \left(\tfrac{2}{3} u^{\frac{3}{2}} \right) + c \\
&= \tfrac{10}{27} (3x^3 + 5)^{\frac{3}{2}} + c
\end{aligned}$$

Summary. Hopefully, you see how completely logical substitution is. We know how the integrand has to look. We just have to get it to look that way.

Make sure that you find du by actually differentiating both sides. Don't just attach a du at the end of the integrand. There has to be a complete substitution.

Example 3: A Substitution with a Square Root in the Denominator

Evaluate
$$\int \frac{(\sqrt{x}+7)^8}{\sqrt{x}}\,dx$$

Strategy. This integral contains the expression $(\sqrt{x}+7)$ to a power. We let $u = \sqrt{x}+7$ as follows.

Solution.
$$\begin{aligned}
\text{Let } u &= \sqrt{x}+7 = x^{\frac{1}{2}}+7 \\
\text{then } du &= \frac{1}{2}x^{-\frac{1}{2}}\,dx \\
\text{or } 2\,du &= \frac{1}{\sqrt{x}}\,dx
\end{aligned}$$

Now we have a quandary. We need to decide whether the substitution can be done, i.e. is there a $\frac{1}{\sqrt{x}}dx$ in the integrand. Let's rewrite the integral as:

$$\begin{aligned}
\int \frac{(\sqrt{x}+7)^8}{\sqrt{x}}\,dx &= \int \underbrace{(\sqrt{x}+7)^8}_{u^8} \underbrace{\frac{1}{\sqrt{x}}\,dx}_{2\,du} \\
&= \int 2u^8\,du \\
&= 2 \cdot \frac{1}{9}u^9 + c \\
&= \frac{2}{9}(\sqrt{x}+7)^9 + c
\end{aligned}$$

Summary. Sometimes when there is a quotient, it helps to rewrite it as a product as shown above. Then it may be easier to decide on the substitution. Note that the substitution in this example worked, because the integral was a product of the inside function $u = \sqrt{x}+7$ raised to some power times twice the derivative of the inside function $2u' = \frac{1}{\sqrt{x}}$.

Example 4: Where the Numerator is the Derivative of the Denominator

Evaluate:
$$\int \frac{6x^2 + 4}{x^3 + 2x} \, dx$$

Strategy. We know how to integrate $\int \frac{1}{u} du$. It is equal to $\ln |u| + c$. Therefore, it is not at all uncommon to let u equal the denominator of the fraction. Let's try it and see what happens.

Solution.
$$\begin{aligned} \text{Let } u &= x^3 + 2x \\ \text{then } du &= \left(3x^2 + 2\right) dx \\ \text{and } 2du &= \left(6x^2 + 4\right) dx \end{aligned}$$

The integral can be rewritten as:

$$\begin{aligned} \int \frac{6x^2 + 4}{x^3 + 2x} dx &= \int \underbrace{\frac{1}{x^3 + 2x}}_{\frac{1}{u}} \cdot \underbrace{\left(6x^2 + 4\right) dx}_{2u\,du} \\ &= 2 \int \frac{1}{u} du \\ &= 2 \ln |u| + c \\ &= 2 \ln \left| x^3 + 2x \right| + c \end{aligned}$$

Summary. As long as du matches the variables in the integrand, the substitution is fine. Any constants by which the integrand is multiplied can be brought in front of the integral sign.

Note that a variable can **never** go in front of the integral sign!

Example 5: A Definite Integral with a Logarithm

Evaluate the definite integral

$$\int_e^{e^4} \frac{dx}{x \ln x}$$

Strategy. Let's write this integral as $\int_e^{e^4} \frac{1}{\ln x} \cdot \frac{1}{x} dx$. We do not have a formula by which to integrate $\ln x$, so that must be what u will stand for. We will see if this substitution takes care of the rest of the integrand.

Solution.

$$\text{Let } u = \ln x$$
$$\text{then } du = \frac{1}{x} dx$$

How nice! We must also substitute for the limits of the integral.

When $x = e$, then $u = \ln e = 1$ and
when $x = e^4$, then $u = \ln e^4 = 4 \ln e = 4$

We now have:

$$\int_e^{e^4} \frac{dx}{x \ln x} = \int_{x=e}^{x=e^4} \underbrace{\frac{1}{\ln x}}_{\frac{1}{u}} \underbrace{\frac{1}{x} dx}_{du}$$
$$= \int_{u=1}^{u=4} \frac{1}{u} du$$
$$= \ln|u| \Big|_1^4$$
$$= \ln 4 - \ln 1 = \ln 4$$

Summary. Note how the u and the du substitute completely for any term involving x. If this does not happen, then the integral cannot be done by substitution. In later sections, we will learn other techniques of integration.

Example 6: A Definite Integral with an Inverse Trigonometric Function

Evaluate
$$\int_0^1 \frac{\sin^{-1} t}{\sqrt{1-t^2}} dt$$

Strategy. In a similar manner to the way we knew to use $u = \ln x$ in example 5, we know here that u must stand for $\sin^{-1} t$. We know that because we have no function whose derivative is $\sin^{-1} t$ and hence no formula for $\int \sin^{-1} t\, dt$. Note also that $\frac{1}{\sqrt{1-t^2}}$ is the derivative of $\sin^{-1} t$, thus the term $\frac{1}{\sqrt{1-t^2}}$ will be part of the du in the u-substitution.

Solution.
$$\text{If } u = \sin^{-1} t$$
$$\text{then } du = \frac{1}{\sqrt{1-t^2}} dt$$

We must also change the limits of integration from t-limits to u-limits. Remember that $\sin^{-1} t$ is the inverse function to $y = \sin t$. The domain for $f(t) = \sin^{-1} t$ is $[-1, 1]$, and the range is $[-\frac{\pi}{2}, \frac{\pi}{2}]$. $\sin^{-1} 0$ asks us for the angle whose sine is 0. $\sin 0 = 0$, but $\sin \pi = 0$, also. Since the angle has to be in the interval $[-\frac{\pi}{2}, \frac{\pi}{2}]$, the answer is 0. Using a similar argument, $\sin^{-1} 1 = \frac{\pi}{2}$, since $\sin \frac{\pi}{2} = 1$, and $\frac{\pi}{2}$ lies in the interval $[-\frac{\pi}{2}, \frac{\pi}{2}]$.

$$\text{If } t = 0, \text{ then } u = \sin^{-1} 0 = 0$$
$$\text{If } t = 1, \text{ then } u = \sin^{-1} 1 = \frac{\pi}{2}$$

The integral can now be rewritten as:

$$\int_0^1 \frac{\sin^{-1} t}{\sqrt{1-t^2}} dt = \int_{t=0}^{t=1} \underbrace{\sin^{-1} t}_{u} \underbrace{\frac{1}{\sqrt{1-t^2}} dt}_{du}$$

$$= \int_{u=0}^{u=\frac{\pi}{2}} u\, du$$

$$= \frac{1}{2} u^2 \Big|_0^{\frac{\pi}{2}}$$

$$= \frac{1}{2}\left(\frac{\pi}{2}\right)^2 - 0 = \frac{\pi^2}{8}$$

Summary. Note that all of the integrals in this section have a pattern. Each integral contains a function g, and also its derivative g'. Here $g(t) = \sin^{-1} t$, and $g'(t) = \frac{1}{\sqrt{1-t^2}}$. The substitution $u = g(t)$ will often work for these integrals.

Example 7: A Substitution that leads to an Expansion of the Integrand

Evaluate
$$\int_1^2 x\sqrt{x-1}\,dx$$

Strategy. We know how to integrate a single letter to a power. Since $\sqrt{x-1} = (x-1)^{\frac{1}{2}}$, we can let $u = x - 1$ as follows.

Solution.
$$\begin{aligned} \text{If } u &= x - 1 \\ \text{then } du &= dx \text{ and} \\ u + 1 &= x \end{aligned}$$

We must also change the limits of integration:

$$\text{If } x = 1, \text{ then } u = 1 - 1 = 0$$
$$\text{If } x = 2, \text{ then } u = 2 - 1 = 1$$

$$\begin{aligned} \int_1^2 \underbrace{x}_{u+1} \underbrace{(x-1)^{\frac{1}{2}}}_{u^{\frac{1}{2}}} \underbrace{dx}_{du} &= \int_0^1 (u+1) \cdot u^{\frac{1}{2}}\,du \\ &= \int_0^1 \left(u^{\frac{3}{2}} + u^{\frac{1}{2}}\right) du \\ &= \left. \frac{2}{5}u^{\frac{5}{2}} + \frac{2}{3}u^{\frac{3}{2}} \right|_0^1 \\ &= \frac{2}{5} + \frac{2}{3} - (0 + 0) \\ &= \frac{16}{15} \end{aligned}$$

Summary. Sometimes if there is an *x leftover* after we find u and du, we realize that substitution will not work. However, sometimes all we have to do is algebraically manipulate what we have and we can cleverly figure it out, as shown in this problem!

4.6. Integration By Parts

Key Topics.
- Integration by Parts
- Solving an Equation for an Unknown Integral
- Reduction Formulas

Worked Examples.
- Integration by Parts
- Integration by Parts Involving a Logarithm
- Repeated Integration by Parts
- Solving for the Unknown Integral
- A Reduction Formula

Overview

All integrals cannot be found by substitution. Another powerful method is called integration by parts. We use this method when the integrand is composed of two parts. For example, it might look like $\int xe^{2x}dx$ where the integrand consists of x, which is an algebraic part, and e^{2x}, which is an exponential part.

Integration by Parts. The formula for integration by parts follows directly from the product rule for differentiation. The actual proof can be found in the text.

It says to separate the integrand into two parts called u and dv. Then

$$\int u\,dv = uv - \int v\,du$$

The part we call u will have to be differentiated to get du. The part called dv will have to be integrated to get v. The hardest part is how to decide which part is u and which part is dv. The acronym **LIATE** will help with that.

L stands for logarithmic function
I stands for inverse trig function
A stands for algebraic function
T stands for trigonometric function
E stands for exponential function.

Whichever letter appears first in LIATE is the function we will call u. For example, in the above $\int xe^{2x}dx$, we have an **a**lgebraic part and an **e**xponential part. Since A is before E in LIATE, we let $u = x$ and $dv = e^{2x}dx$. This will be more understandable as we work the example problems.

Solving an Equation for an Unknown Integral. Sometimes when we use parts, it seems as though we are going around in circles. Example 4 coming up, is like that. When we get there we will see how to complete the problem.

Reduction Formulas. Sometimes we start with a function to a power. As we proceed by parts, the power continually decreases until it gets to zero. Example 5 will show how to complete a problem like that.

Example 1: Integration by Parts

$$\int xe^x\,dx$$

Strategy. As stated before, we have an algebraic part, namely x, and an exponential part, $e^x dx$. Since A, for algebraic, comes before E, for exponential, in LIATE, we let $u = x$. Then the rest of the integrand $e^x dx$ must be dv.

Solution. We let $u = x$, and $dv = e^x dx$. To obtain du we differentiate x, to obtain v we integrate both sides of $dv = e^x dx$.

$$u = x \qquad dv = e^x dx$$
$$du = dx \qquad v = e^x$$

Then

$$\int \underbrace{x}_{u}\underbrace{e^x dx}_{dv} = \underbrace{x}_{u}\underbrace{e^x}_{v} - \int \underbrace{e^x}_{v}\underbrace{dx}_{du}$$
$$= xe^x - e^x + c$$

Summary. It is essential to have udv contain the entire integrand. This includes the dx. The dx must be there so that the expression can be integrated. In this example, we had

$$dv = e^x dx$$
$$\int dv = \int e^x dx$$
$$v = e^x$$

We do not need $+c$ for this step when using integration by parts.

Example 2: Integration by Parts Involving a Logarithm

$$\int x^2 \ln x \, dx$$

Strategy. In this problem we have an algebraic part, x^2, and a logarithmic part, $\ln x$. Since L, for logarithm, precedes A, for algebraic, in LIATE, we let $u = \ln x$, and then we need the rest of the integrand to be dv. Consequently, we have $dv = x^2 dx$.

Solution.

$$u = \ln x \qquad dv = x^2 dx$$
$$du = \frac{1}{x} dx \qquad v = \tfrac{1}{3} x^3$$

$$\begin{aligned}
\int x^2 \ln x \, dx &= \int \underbrace{\ln x}_{u} \underbrace{x^2 dx}_{dv} = \underbrace{\ln x}_{u} \underbrace{\tfrac{1}{3} x^3}_{v} - \int \underbrace{\tfrac{1}{3} x^3}_{v} \underbrace{\tfrac{1}{x} dx}_{du} \\
&= \tfrac{1}{3} x^3 \ln x - \int \tfrac{1}{3} x^2 dx \\
&= \tfrac{1}{3} x^3 \ln x - \tfrac{1}{9} x^3 + c
\end{aligned}$$

Summary. Note that once again when we chose u and dv we have to choose the terms of the **entire integral**, that is $\int u \, dv =$ entire integral.

Example 3: Repeated Integration by Parts

$$\int_0^1 x^2 e^x \, dx$$

Strategy. In this problem we need to use integration by parts twice. We are working with algebraic and exponential parts. The algebraic part will be u in both cases.

Solution.
$$u = x^2 \qquad dv = e^x dx$$
$$du = 2x\,dx \qquad v = e^x$$

$$\int_0^1 \underbrace{x^2}_{u} \underbrace{e^x dx}_{dv} = \underbrace{x^2}_{u} \underbrace{e^x}_{v} \Big|_0^1 - \int_0^1 \underbrace{e^x}_{v} \underbrace{2x\,dx}_{du}$$
$$= (e - 0) - \int_0^1 2xe^x\,dx$$

We use integration by parts once more in order to integrate $\int_0^1 2xe^x\,dx$.

$$u = 2x \qquad dv = e^x dx$$
$$du = 2\,dx \qquad v = e^x$$

$$\int_0^1 2xe^x\,dx = 2xe^x\big|_0^1 - \int_0^1 2e^x\,dx$$
$$= 2xe^x - 2e^x\big|_0^1$$
$$= (2e - 2e) - (0 - 2) = 2$$

Putting it all together, we have:

$$\int_0^1 x^2 e^x\,dx = (e - 0) - \int_0^1 2xe^x\,dx$$
$$= e - 2$$

Summary. When we used integration by parts the first time, we ended up with another integral which needed to be done by parts also. It is important, and sometimes difficult, to keep track of all the terms, especially the signs, when we use parts more than once in a single problem. The next example shows this quite well.

Example 4: Solving for the Unknown Integral

$$\int e^x \cos x \, dx$$

Strategy. We have here a trigonometric function and an exponential function. We let $u = \cos x$, and $dv = e^x dx$. However, it is obvious that no matter how many times we integrate e^x, it will still be e^x. Also, as we differentiate $\cos x$ we will go up and back between $\cos x$ and $\sin x$ as our function. The signs will vary, but basically we keep getting $\sin x$ or $\cos x$ with e^x. We will use integration by parts twice and watch what happens.

Solution. Start with integration by parts.

$$u = \cos x \qquad dv = e^x dx$$
$$du = -\sin x \, dx \qquad v = e^x$$

$$\begin{aligned}\int e^x \cos x \, dx &= e^x \cos x - \int e^x (-\sin x) \, dx \\ &= e^x \cos x + \int e^x \sin x \, dx\end{aligned}$$

Use integration by parts again.

$$u = \sin x \qquad dv = e^x dx$$
$$du = \cos x \, dx \qquad v = e^x$$

Thus

$$\int e^x \sin x \, dx = e^x \sin x - \int e^x \cos x \, dx$$

$$\implies \int e^x \cos x \, dx = e^x \cos x + e^x \sin x - \int e^x \cos x \, dx$$

Look carefully! We are trying to find $\int e^x \cos x \, dx$ which now occurs on both sides of the equation. Whenever we are solving an equation and the variable appears on both sides of the equal sign, we get all the variables on the same side. Let $I = \int e^x \cos x \, dx$, just to make the computation a little less messy.

$$\begin{aligned}I &= e^x \cos x + e^x \sin x - I \\ 2I &= e^x \cos x + e^x \sin x \\ I &= \frac{1}{2}(e^x \cos x + e^x \sin x) + c \\ \int e^x \cos x \, dx &= \frac{1}{2}(e^x \cos x + e^x \sin x) + c\end{aligned}$$

Summary. The technique used in this problem is pretty clever! Whenever it looks like you keep bouncing up and back between functions, look to see if the same integral appears on both sides of the equal sign. Then algebraically get both integrals on the same side and divide by the coefficient. If it is an indefinite integral, it needs a $+c$ at the very end.

Example 5: A Reduction Formula

Use integration by parts to prove for any positive integer n that

$$\int (\ln x)^n \, dx = x (\ln x)^n - n \int (\ln x)^{n-1} \, dx$$

Use this to evaluate $\int (\ln x)^3 \, dx$.

Strategy. The two parts here are logarithmic $((\ln x)^n)$ and algebraic (dx). It is called a reduction formula, because the power reduces by 1 each time it is used.

Solution.

$$u = (\ln x)^n \qquad dv = dx$$
$$du = n (\ln x)^{n-1} \frac{1}{x} dx \qquad v = x$$

$$\int (\ln x)^n \, dx = x (\ln x)^n - \int xn (\ln x)^{n-1} \frac{1}{x} dx$$
$$= x (\ln x)^n - n \int (\ln x)^{n-1} \, dx$$

To evaluate $\int (\ln x)^3 dx$ we use this formula starting with $n = 3$,

$$\int (\ln x)^3 \, dx = x (\ln x)^3 - 3 \int (\ln x)^2 \, dx$$

We can use the formula again for $n = 2$,

$$\int (\ln x)^2 \, dx = x(\ln x)^2 - 2 \int \ln x \, dx$$

and thus

$$\int (\ln x)^3 dx = x (\ln x)^3 - 3 \left[x(\ln x)^2 - 2 \int \ln x \, dx \right] = x(\ln x)^3 - 3x(\ln x)^2 + 6 \int \ln x \, dx$$

Using the formula one more time with $n = 1$,

$$\int \ln x \, dx = x \ln x - 1 \int dx$$

And finally

$$\int (\ln x)^3 \, dx = x (\ln x)^3 - 3x(\ln x)^2 + 6 \left[x \ln x - 1 \int dx \right]$$
$$= x (\ln x)^3 - 3x(\ln x)^2 + 6x \ln x - 6x + c$$

Summary. You can see in this problem how extremely careful you need to be when there are so many numbers, variables and signs all over the place. Make use of parentheses and brackets to help.

4.7. Other Techniques of Integration

Key Topics.
- Partial Fractions
- Trigonometric Integrals
- Trigonometric Substitutions

Worked Examples.
- Setting up Partial Fractions
- Partial Fractions with Distinct Factors
- Partial Fraction with a Quadratic Factor
- Trigonometric Integrals with Odd Powers of Sine or Cosine
- Even Powers of Sine and Cosine
- An Example Involving Tangent and Secant
- A Trigonometric Substitution

Overview

We have learned about integration by substitution and by parts. What about problems which are not integrable using these methods? This section covers several other methods of integration. Deciding which method to use is one of the many challenges of integration.

Partial Fractions. We've done many algebraic problems where we have found a common denominator and then added or subtracted rational expressions. In this section we take the *answer* and decompose it back into the original rational expressions.

Trigonometric Integrals. Here we deal with integrals containing trigonometric functions in the integrand. The Pythagorean identities are used extensively in this type of problem. The ones you need to be familiar with are:

$$\sin^2\theta + \cos^2\theta = 1$$
$$1 + \tan^2\theta = \sec^2\theta$$
$$1 + \cot^2\theta = \csc^2\theta$$

Also needed here are the trigonometric derivatives.

Trigonometric Substitutions. The Pythagorean Theorem is used extensively for trigonometric substitution. So far, we only used u-substitutions. Here we will learn when and how to use the trig functions, $\sin\theta$, $\tan\theta$, and $\sec\theta$, for a new type of substitution.

Example 1: Setting up Partial Fractions

Set up the partial fraction decompositions for the following functions.

a) $\dfrac{10x - 11}{x^2 - 2x - 3}$

b) $\dfrac{10x^2 + 5}{(x - 7)^2 (x - 1)}$

c) $\dfrac{10x^3 - 11}{(x^2 - 1)(x^2 + 1)}$

Strategy. We will analyze the denominators in these problems. First, they must be fully factored. The three types of denominators that we can have if we *factor* the denominator are linear factors, repeated linear factors, and quadratic factors.

Solution. a) First factor the denominator into two linear factors.
$$\frac{10x - 11}{x^2 - 2x - 3} = \frac{10x - 11}{(x-3)(x+1)} = \frac{A}{x-3} + \frac{B}{x+1}$$

b) This rational expression is already fully factored. The $(x+7)^2$ is what we call a repeated linear factor. We need to use the denominator $(x+7)$ as well as the denominator $(x+7)^2$.
$$\frac{10x^2 + 5}{(x-7)^2(x-1)} = \frac{A}{(x-7)} + \frac{B}{(x-7)^2} + \frac{C}{(x-1)}$$

c) In this last example, we can factor the denominator. Then we will have two linear factors and one quadratic factor.
$$\frac{10x^3 - 11}{(x^2-1)(x^2+1)} = \frac{10x^3 - 11}{(x-1)(x+1)(x^2+1)}$$
$$= \frac{A}{(x-1)} + \frac{B}{(x+1)} + \frac{Cx+D}{(x^2+1)}$$

Summary. You need to learn the pattern for decomposing each type of rational expression. In the next example, we will solve for the constants A, B, C, and D, which will make it possible (even easy) for us to carry out the integration. Keep in mind that we can only apply partial fractions to a rational function if the degree of the denominator is strictly larger than the degree of the numerator as in the examples above. If the degree of the numerator is equal or larger than the degree of the denominator, for example $\dfrac{x^2+1}{x^2-1}$, or $\dfrac{x^4 - 3x + 10}{x^3 - x}$, then you first have to perform a polynomial division.

Example 2: Partial Fractions with Distinct Factors

Evaluate
$$\int \frac{(5x-3)\,dx}{(x-1)(x-2)(x-3)}$$

Strategy. Luckily, this denominator is already fully factored into three linear factors. We use partial fraction decomposition in order to rewrite this expression in a form that we can integrate.

Solution.
$$\frac{(5x-3)}{(x-1)(x-2)(x-3)} = \frac{A}{x-1} + \frac{B}{x-2} + \frac{C}{x-3}$$

Multiply both sides of the equation by the common denominator: $(x-1)(x-2)(x-3)$

$$5x-3 = A(x-2)(x-3) + B(x-1)(x-3) + C(x-1)(x-2)$$

This equation holds for all x. In particular if we choose $x=1$, $x=2$, or $x=3$, then one of the factors is zero, making it possible to solve for A, B, and C.
If $x=1$, then:
$$5(1) - 3 = A(1-2)(1-3) + B(0)(1-3) + C(0)(1-2)$$
$$2 = 2A \implies A = 1$$

Using the same method, let $x = 2$,

$$7 = 0 + B(1)(-1) + 0$$
$$7 = -B \implies B = -7$$

Lastly, if $x = 3$ then

$$12 = 0 + 0 + C(2)(1)$$
$$12 = 2C \implies C = 6$$

Therefore, we can rewrite:

$$\int \frac{(5x-3)\,dx}{(x-1)(x-2)(x-3)} = \int \frac{1}{x-1}dx + \int \frac{-7}{x-2}dx + \int \frac{6}{x-3}dx$$
$$= \ln|x-1| - 7\ln|x-2| + 6\ln|x-3| + c$$

Summary. As you can see, once the expression is decomposed into *simpler* rational expressions, the integration is more straightforward. In the next example, we'll see what happens with a quadratic factor.

Example 3: Partial Fraction with a Quadratic Factor

Evaluate
$$\int \frac{2x^2 - 5x + 2}{x^3 + x} dx$$

Strategy. This denominator factors into a linear factor and a quadratic factor. Because of this different set-up, we will use a slightly different method of finding the constants, though the method from the previous example does work for this example, also. If both sides of the equation are completely multiplied out, then we can equate the coefficients of like powers of x.

Solution.
$$\frac{2x^2 - 5x + 2}{x^3 + x} = \frac{2x^2 - 5x + 2}{x(x^2 + 1)} = \frac{A}{x} + \frac{Bx + C}{x^2 + 1}$$

Multiply both sides of the equation by the common denominator $x(x^2 + 1)$ and get

$$\begin{aligned} 2x^2 - 5x + 2 &= A(x^2 + 1) + (Bx + C)x \\ 2x^2 - 5x + 2 &= Ax^2 + A + Bx^2 + Cx \\ 2x^2 - 5x + 2 &= (A + B)x^2 + Cx + A \end{aligned}$$

The coefficient of x^2 on the left is 2 and on the right is $(A+B)$. Hence, $A+B = 2$. Similarly, $-5 = C$ and $2 = A$. Substituting 2 for A, we find $2 = 2 + B$ which says that $B = 0$.

Therefore, we can write: $\quad \dfrac{2x^2 - 5x + 2}{x(x^2 + 1)} = \dfrac{2}{x} + \dfrac{-5}{x^2 + 1}.$

Now we can integrate:

$$\begin{aligned} \int \frac{2x^2 - 5x + 2}{x^3 + x} dx &= \int \frac{2dx}{x} + \int \frac{-5dx}{x^2 + 1} \\ &= 2\ln|x| - 5\tan^{-1} x + c \end{aligned}$$

Summary. If there is a rational expression to integrate and the denominator is factorable, then one method to try is partial fraction decomposition.

Example 4: Trigonometric Integrals with Odd Powers of Sine or Cosine

Evaluate the following trigonometric integrals:

a) $\quad \int \sin^4 x \cos x\, dx$

b) $\quad \int \cos^5 x\, dx$

Strategy. For part a) we have $\sin x$ and $\cos x$ which are basically the derivatives of each other. We know how to integrate u^4, so let $u = \sin x$.

In part b) we will rewrite $\cos^5 x$ as a product of $\cos x$ and an expression involving copies of $\sin x$, but no $\cos x$. For this we can use the Pythagorean identity,
$$\cos^2 x = 1 - \sin^2 x$$
Then we use the property that the derivative of $\sin x$ is $\cos x$, again let $u = \sin x$.

Solution. a) Use the substitution $u = \sin x$, $du = \cos x\, dx$. Then
$$\int \underbrace{\sin^4 x}_{u^4} \underbrace{\cos x\, dx}_{du} = \int u^4\, du = \frac{1}{5} u^5 + c = \frac{1}{5} \sin^5 x + c$$

b) We first rewrite $\cos^5 x$:
$$\begin{aligned} \cos^5 x &= \cos^4 x \cdot \cos x \\ &= (\cos^2 x)^2 \cdot \cos x \\ &\, (1 - \sin^2 x)^2 \cdot \cos x \end{aligned}$$

Then let $u = \sin x$, $du = \cos x\, dx$:
$$\begin{aligned} \int \cos^5 x\, dx &= \int \underbrace{(1 - \sin^2 x)^2}_{(1-u^2)^2} \underbrace{\cos x\, dx}_{du} \\ &= \int (1 - u^2)^2\, du \\ &= \int (1 - 2u^2 + u^4)\, du \\ &= u - \frac{2}{3} u^3 + \frac{1}{5} u^5 + c \\ &= \sin x - \sin^3 x + \frac{1}{5} \sin^5 x + c \end{aligned}$$

Summary. In this example, both integrals contained odd powers of cosine. When an integral contains odd powers of cosine, and odd or even powers of sine, then the trick is to *pull out* one copy of cosine, and rewrite the rest in terms of sine with the help of the Pythagorean identities. Then the integral can be evaluated by using the substitution $u = \sin x$, $du = \cos x\, dx$. The same trick works with odd powers of sine by *pulling out* a copy of sine and rewriting the rest in terms of cosine. In this case the substitution is $u = \cos x$, $du = -\sin x$.

Example 5: Even Powers of Sine and Cosine

Evaluate
$$\int \sin^2 x \cos^2 x \, dx$$

Strategy. When both sine and cosine are raised to an even power, we have to be even more clever in how we rearrange and change the integrand so that we can integrate. We use the half-angle formulas:

$$\sin^2 \theta = \frac{1 - \cos 2\theta}{2}$$
$$\cos^2 \theta = \frac{1 + \cos 2\theta}{2}$$

Solution. Let's work with the integrand and get it into a form which we can integrate. Using the half-angle formulas:

$$(\sin^2 x)(\cos^2 x) = \left(\frac{1 - \cos(2x)}{2}\right)\left(\frac{1 + \cos(2x)}{2}\right)$$
$$= \frac{1 - \cos^2(2x)}{4} = \frac{1}{4} - \frac{1}{4}\cos^2(2x)$$

Note that we still have $\cos^2 2x$, which cannot be integrated as it stands. We need to use the half-angle formula again. We must be extremely careful here. The formula says that if the given angle is θ, then the numerator contains 2θ. Here we have the angle of $2x$. If $\theta = 2x$, then $2\theta = 4x$. Consequently,

$$\cos^2(2x) = \frac{1 + \cos(4x)}{2} = \frac{1}{2} + \frac{1}{2}\cos(4x)$$

Now we can integrate the entire expression:

$$\int \sin^2 x \cos^2 dx = \int \left(\frac{1}{4} - \frac{1}{4}\left(\frac{1}{2} + \frac{1}{2}\cos(4x)\right)\right) dx$$
$$= \int \left(\frac{1}{4} - \frac{1}{8} - \frac{1}{8}\cos(4x)\right) dx$$
$$= \int \left(\frac{1}{8} - \frac{1}{8}\cos(4x)\right) dx$$
$$= \frac{1}{8}x - \frac{1}{8} \cdot \left(\frac{1}{4}\sin(4x)\right) = \frac{1}{8}x - \frac{1}{32}\sin(4x) + c$$

Summary. The half-angle formulas are often used when we only have even powers of sine and cosine, here $\sin^2 x$ and $\cos^2 x$. In this example after applying half-angle formulas twice, we end up with $\cos(4x)$, a function that we can integrate!

Example 6: An Example Involving Tangent and Secant

Evaluate
$$\int \frac{\sec^2 x}{\tan^2 x} dx$$

Strategy. Recognize that the derivative of $\tan x$ is $\sec^2 x$ which is in the numerator. Thus we can use the u-substitution $u = \tan x$.

Solution. If $u = \tan x$, then $du = \sec^2 x\, dx$
$$\int \frac{\sec^2 x}{\tan^2 x} dx = \int \frac{du}{u^2} = \int u^{-2} du = -u^{-1} + c = -\frac{1}{\tan x} + c$$

Summary. If an integrand contains secants and tangents, then look for a substitution using either the fact that $\frac{d}{dx}(\tan x) = \sec^2 x$, or the fact that $\frac{d}{dx}(\sec x) = \sec x \tan x$.

The secant and tangent derivatives are intermixed as are the sine and cosine derivatives, and the cotangent and cosecant derivatives too.

Example 7: A Trigonometric Substitution

Use the trigonometric substitution $x = 3\sin\theta$ to evaluate the integral

$$\int \frac{x^3\,dx}{\sqrt{9-x^2}}$$

Strategy. $\sqrt{9-x^2}$ leads us to think of the Pythagorean Theorem. In a right triangle, if the hypotenuse is 3 and the side opposite angle θ is x, then the other leg is $\sqrt{9-x^2}$ as shown below. We will use this triangle as a *reference triangle* to find the sine, cosine, tangent, or other trig function of θ.

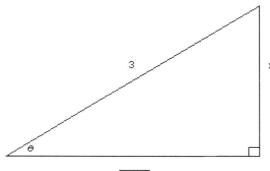

We then have $\sin\theta = \dfrac{x}{3}$, from which we get our substitution,

$$\text{let } x = 3\sin\theta$$
$$\text{then } dx = 3\cos\theta\,d\theta$$

Consequently,

$$\begin{aligned}
\sqrt{9-x^2} &= \sqrt{9-(3\sin\theta)^2} \\
&= \sqrt{9-9\sin^2\theta} \\
&= \sqrt{9(1-\sin^2\theta)} \\
&= 3\sqrt{\cos^2\theta} \\
&= 3\cos\theta
\end{aligned}$$

Alternatively, we can get this information directly from the reference triangle. $\cos\theta = \dfrac{\text{adjacent side}}{\text{hypotenuse}} = \dfrac{\sqrt{9-x^2}}{3}$, thus $3\cos\theta = \sqrt{9-x^2}$.

Solution.

4.7. OTHER TECHNIQUES OF INTEGRATION

$$\int \frac{\overbrace{x^3}^{27\sin^3\theta}\,\overbrace{dx}^{3\cos\theta d\theta}}{\underbrace{\sqrt{9-x^2}}_{3\cos\theta}} = \int \frac{(27\sin^3\theta)(3\cos\theta d\theta)}{3\cos\theta}$$

$$= \int 27\sin^3\theta d\theta$$

(a trig integral with an odd power of sine)

$$= \int 27\left(\sin^2\theta\right)\sin\theta d\theta$$

$$= \int 27\left(1-\cos^2\theta\right)\sin\theta d\theta$$

(use the substitution $u = \cos\theta, du = -\sin\theta d\theta$)

$$= -\int 27\left(1-u^2\right)du$$

$$= -27u + \frac{27}{3}u^3 + c$$

$$= -27\cos\theta + 9\cos^3\theta + c$$

Since the given integral is in terms of x, our answer must also be in terms of x. Look back at the triangle and see that $\cos\theta = \frac{\sqrt{9-x^2}}{3}$. Thus

$$\int \frac{x^3 dx}{\sqrt{9-x^2}} = -27\left(\frac{\sqrt{9-x^2}}{3}\right) + 9\left(\frac{\sqrt{9-x^2}}{3}\right)^3 + c$$

$$= -9\sqrt{9-x^2} + \frac{9}{27}\sqrt{(9-x^2)^3} + c$$

Note that $\sqrt{(9-x^2)^3} = \sqrt{(9-x^2)^2} \cdot \sqrt{(9-x^2)} = (9-x^2)\cdot\sqrt{9-x^2}$. Now we can factor out the common factor of $\sqrt{9-x^2}$ and simplify.

$$\int \frac{x^3 dx}{\sqrt{9-x^2}} = -9\sqrt{9-x^2} + \frac{1}{3}(9-x^2)\sqrt{9-x^2} + c$$

$$= \sqrt{9-x^2}\left[-9 + \frac{1}{3}(9-x^2)\right] + c$$

$$= \sqrt{9-x^2}\left[-9 + 3 - \frac{1}{3}x^2\right] + c$$

$$= \sqrt{9-x^2}\left[-6 - \frac{1}{3}x^2\right] + c$$

Summary. An interesting problem. In a trigonometric substitution, we always end up with a trigonometric integral. Often you will have to use another substitution, in this example $u = \cos\theta$, to solve the integral. Make sure your final answers is in terms of the original variable. Drawing a *reference triangle* can help a lot to find the values of the trigonometric functions that we get from the trigonometric integral.

You can draw *reference triangles* for other trigonometric substitutions, also. For example for the trig substitution $x = 2\tan\theta$, we know that $\tan\theta = \dfrac{x}{2} = \dfrac{\text{opposite side}}{\text{adjacent side}}$, and the corresponding *reference triangle* will be a right triangle with opposite side $= x$, adjacent side $= 2$, and hypotenuse $= \sqrt{x^2 + 4}$.

4.8. Integration Tables

Key Topics.
- Using Tables of Integrals

Worked Examples.
- Making a Substitution Before Using a Table
- Completing the Square Before Using an Integral Table

Overview

The textbook contains a table of integrals in the back of the book. Most calculus texts have a table, some bigger and some smaller.

Using Tables of Integrals. Unfortunately, this is often harder than it first appears. When looking up an integral in the table, it must be in **exactly the same form** as it is in the table. Usually, a, b, and c are constants, while u and x are variables. If the integrand in the table has u^2, then our integrand must have a single letter squared. This means, for example, that $4x^2$ is not good enough. $(2x^2)$ is also not good enough. As seen in the example problem we need to substitute $u = 2x$ and then have u^2 before any integration takes place. Once we have identified which entry in the table to use, and transformed the form of our integral to look like the one in the table, then the rest is easy. We just put the constants and variables exactly where the table tells us to put them.

Following are two examples showing common algebraic manipulations and substitutions to make before using a table of integrals.

Example 1: Making a Substitution Before Using a Table

Evaluate
$$\int \frac{x^2\,dx}{\sqrt{16x^2+9}}$$

Strategy. We look at the table of integration in the back of the text, and we look for the section which has formulas involving $\sqrt{u^2+a^2}$. Then we look for one which has a fraction and also has a squared variable in the numerator. Entry #26 looks promising.

Entry #26 : $\int \dfrac{u^2}{\sqrt{a^2+u^2}}\,du = \dfrac{1}{2}u\sqrt{a^2+u^2} - \dfrac{1}{2}a^2 \ln\left|u+\sqrt{a^2+u^2}\right| + c$

Solution. The difficulty is that the table entry has a^2+u^2 in the denominator. We must make our denominator look like that one. $16x^2 = (4x)^2$, which is not a single letter squared. Therefore, we need to make a substitution:

$$\begin{aligned} \text{If } u &= 4x \\ \text{then } du &= 4dx \\ \text{and } \tfrac{1}{4}du &= dx \end{aligned}$$

$$\begin{aligned}
\int \frac{x^2}{\sqrt{16x^2+9}}\,dx &\\
&= \frac{1}{4}\int \frac{\left(\frac{u}{4}\right)^2}{\sqrt{u^2+9}}\,du \\
&= \frac{1}{64}\int \frac{u^2}{\sqrt{u^2+9}}\,du && (a=3, \text{use entry } \#26) \\
&= \frac{1}{64}\left(\frac{1}{2}u\sqrt{9+u^2} - \frac{1}{2}3^2 \ln\left|u+\sqrt{9+u^2}\right|\right) + c \\
&= \frac{1}{128}\cdot 4x\sqrt{9+16x^2} - \frac{9}{2}\cdot\frac{1}{64}\ln\left|4x+\sqrt{9+16x^2}\right| + c \\
&= \frac{1}{32}x\sqrt{9+16x^2} - \frac{9}{128}\ln\left|4x+\sqrt{9+16x^2}\right| + c
\end{aligned}$$

Summary. As you can see, the first challenge is to identify the correct entry in the table. Then the second challenge is to get the integrand in the problem to be in the exact same form as the integrand in the table.

Example 2: Completing the Square Before Using an Integral Table

Evaluate
$$\int \frac{2dx}{\sqrt{x^2 + 6x + 13}}$$

Strategy. There are no table entries with the square root of a trinomial. Many entries have squares. Therefore, it seems reasonable to complete the square in the denominator. It also may make things easier if we bring the 2 out in front of the integral sign.

Solution. We first complete the square.
$$x^2 + 6x + 13 = \left(x^2 + 6x + 9\right) + 4 = (x+3)^2 + 4$$
Substitute $u = x + 3$, $du = dx$
$$\int \frac{2dx}{\sqrt{x^2 + 6x + 13}} = 2\int \frac{dx}{\sqrt{(x+3)^2 + 4}}$$
$$= 2\int \frac{du}{\sqrt{u^2 + 4}}$$

Now the integral is in a form where we can search for a table entry to use. Look in the table for an entry which has a 1 in the numerator and $\sqrt{u^2 + a^2}$ in the denominator.

$$\text{Entry } \#25: \int \frac{du}{\sqrt{a^2 + u^2}} = \ln\left|u + \sqrt{a^2 + u^2}\right| + c$$

We have $a = 2$, so the integral is equal to
$$= 2\ln\left|u + \sqrt{4 + u^2}\right| + c$$
$$= 2\ln\left|x + 3 + \sqrt{4 + (x+3)^2}\right| + c$$
$$= 2\ln\left|x + 3 + \sqrt{x^2 + 6x + 13}\right| + c$$

Summary. Completing the square is a very common technique when using integral tables. As you see here, it allows the integrands to be put into a form with squares that match the form of the integrands in the tables.

4.9. Numerical Integration

Key Topics.
- Trapezoidal Rule
- Simpson's Rule
- Error Bounds

Worked Examples.
- Using Trapezoidal and Simpson's Rule for a Function Given by a Table of Values
- Using Trapezoidal and Simpson's Rule for a Function Given by a Formula
- Determine Steps Needed for an Accuracy of 10^{-7}

Overview

For integrals without simple antiderivatives, we have three methods of numerical integration which we can use. We have already seen Riemann sums. If we choose the midpoint of each interval to evaluate the height of each rectangle, then we call it the midpoint rule. Following are two other methods which increase the accuracy of the approximation.

Trapezoidal Rule. Instead of drawing rectangles between the graph and the x-axis, we can draw a slanted line at the top which more closely follows the curve. In that way, we have trapezoids and we can use the area for trapezoids. If we have a regular partition of the intervals, with n subintervals of length $\Delta x = \frac{b-a}{n}$, then

$$T_n(f) = \frac{b-a}{2n}[f(x_0) + 2f(x_1) + 2f(x_2) + ... + 2f(x_{n-1}) + f(x_n)]$$

Simpson's Rule. To obtain an even more accurate approximation, we can use Simpson's Rule. Here we once again construct a regular partition of the interval $[a,b]$ into an **even** number, n, of subintervals. Instead of connecting each pair of points, we connect 3 consecutive points with a parabolic segment. Then

$$S_n(f) = \frac{b-a}{3n}[f(x_0) + 4f(x_1) + 2f(x_2) + 4f(x_3) + 2f(x_4) + ... + 4f(x_{n-1}) + f(x_n)]$$

Error Bounds. Clearly, each of these methods are approximations of $\int_a^b f(x)dx$. We need to know how large of an error it is possible to make. Using ET_n and ES_n to stand for the error using the Trapezoidal Rule and the error using Simpson's Rule, respectively, for $|f''(x)| \leq K$, and $|f^{(4)}(x)| \leq L$, for all x in $[a,b]$ we have,

$$|ET_n| \leq \frac{K(b-a)^3}{12n^2}$$

$$|ES_n| \leq \frac{L(b-a)^5}{180n^4}$$

Example 1: Using Trapezoidal and Simpson's Rule for a Function Given by a Table of Values

Snow is falling over a three hour period at the following rates.

t (in hours)	0	0.5	1	1.5	2	2.5	3
r (in inches per hour)	1.0	0.5	1.5	1.5	1.0	0.5	0

Use the Trapezoidal Rule and Simpson's Rule to approximate the total amount of snow that fell during this period.

Strategy. There are $n = 6$ intervals with each interval having length 0.5. All we need to do is put the numbers from the table into both the Trapezoidal Rule and Simpson's Rule.

Note that from the table we have $f(0) = 1$, $f(0.5) = 0.5$, $f(1) = 1.5$ etc.

Solution. $n = 6$, length of the subinterval $\Delta t = 0.5$.

$$T_6 = \frac{3-0}{2(6)} (1.0 + 2 \cdot 0.5 + 2 \cdot 1.5 + 2 \cdot 1.5 + 2 \cdot 1.0 + 2 \cdot 0.5 + 0)$$
$$= \frac{1}{4} \cdot 11 = \frac{11}{4}$$

$$S_6 = \frac{3-0}{3(6)} (1.0 + 4 \cdot 0.5 + 2 \cdot 1.5 + 4 \cdot 1.5 + 2 \cdot 1.0 + 4 \cdot 0.5 + 0)$$
$$= \frac{1}{6} \cdot 16 = \frac{8}{3}$$

Summary. With all the values set up nicely in a table, all that is left for us to do is plug the numbers into the correct places following the formulas exactly.

Example 2: Using Trapezoidal and Simpson's Rule for a Function Given by a Formula

Use Trapezoidal and Simpson's Rule with $n = 4$ to approximate

$$\int_0^4 \frac{120}{x+1} dx$$

Strategy. If we use $n = 4$ to partition the interval $[0, 4]$, then the x_i's will occur at $0, 1, 2, 3,$ and 4. Note there are five x_i's making 4 subintervals. It will help to create a table of all the values so that we can proceed as we did in example 1.

Solution. We first create a table of values, $n = 4$, $\Delta x = 1$.

x	0	1	2	3	4
$f(x)$	120	60	40	30	24

$$T_4 = \frac{4-0}{2(4)} (120 + 2 \cdot 60 + 2 \cdot 40 + 2 \cdot 30 + 24)$$

$$= \frac{1}{2} \cdot 404 = 202$$

$$S_4 = \frac{4-0}{3(4)} (120 + 4 \cdot 60 + 2 \cdot 40 + 4 \cdot 30 + 24)$$

$$= \frac{1}{3} \cdot 584 = 194.\overline{6}$$

Summary. Once again you can see that once you have a table of values, it is not difficult to plug in the numbers. The difficulty is that to better the approximation we really need many more intervals which makes the calculations exceedingly tedious. A computer or calculator needs to be used.

4.9. NUMERICAL INTEGRATION

Example 3: Determine Steps Needed for an Accuracy of 10^{-7}

Determine how many steps are needed to estimate $\int_0^{10} \sin x \, dx$ to guarantee an accuracy of 10^{-7} using a) Trapezoidal Rule and b) Simpson's Rule.

Strategy. The first thing we need to do in each case is find an upper bound for the respective derivatives needed for each rule. An important thing to remember whenever you are looking for a bound for either the sine function or the cosine function is that $|\sin \theta| \leq 1$ and $|\cos \theta| \leq 1$.

Solution. a) We first find a constant K for trapezoidal approximations.

$$\begin{aligned} f(x) &= \sin x, \; f'(x) = \cos x, \; f''(x) = -\sin x \\ |f''(x)| &= |-\sin x| = |\sin x| \leq 1 \text{ for all } x \\ &\implies K = 1 \end{aligned}$$

Using the error approximation, we get

$(K=1, b-a=10)$ $\qquad |E_T| \leq \dfrac{K(b-a)^3}{12n^2} = \dfrac{10^3}{12n^2}$

Since we want the error to be smaller than 10^{-7}, we need to solve the inequality

$$\begin{aligned} \dfrac{10^3}{12n^2} &\leq 10^{-7} & \text{(multiply by } 10^7 n^2 \text{)} \\ \dfrac{10^{10}}{12} &\leq n^2 & \text{(take square root)} \\ n &\geq 28867.51\ldots \end{aligned}$$

Thus $n \geq 28868$ guarantees an accuracy of 10^{-7}.

b) We first find a constant L for Simpson's Rule approximations.

$$\begin{aligned} f'''(x) &= -\cos x, \; f^{(4)}(x) = \sin x \\ \left|f^{(4)}(x)\right| &= |\sin x| \leq 1 \text{ for all } x \\ &\implies L = 1 \end{aligned}$$

Using the error approximation, we get

$(L=1, b-a=10)$ $\qquad |E_s| \leq \dfrac{L(b-a)^5}{180n^4} = \dfrac{10^5}{180n^4}$

Since we want the error to be smaller than 10^{-7}, we need to solve the inequality

$$\dfrac{10^5}{180n^4} \leq 10^{-7} \qquad \text{(multiply by } 10^7 n^4\text{)}$$

$$\frac{10^{12}}{180} \leq n^4 \qquad \text{(take 4th root)}$$
$$n \geq 273.01...$$

Thus $n \geq 274$ guarantees an accuracy of 10^{-7}.

Summary. The hardest part in this type of problem is usually finding the upper bound. However, with the functions $\sin\theta$ and $\cos\theta$, we always have an upper bound of 1. From there it is just a matter of finding the needed derivatives and substituting the numbers into the error formula.

4.10. Improper Integrals

Key Topics.
- Improper Integrals
- Comparison Test

Worked Examples.
- Examples of Integrals with Discontinuous Integrands
- Integrals with Infinite Limits
- An Example Using Integration by Parts and L'Hôpital's Rule
- Using the Comparison Test to Show that an Integral Converges
- Using the Comparison Test to Show that an Integral Diverges

Overview

In this section we discuss integrals that are not continuous on the interval of integration, and integrals that have $\pm\infty$ for one or both limits of integration.

Improper Integrals. When we do not have finite numbers for the limits of integration or the interval of integration contains any discontinuities in the integrand, then we have an improper integral. We replace the *problem spots* with a variable and then take the limit as that variable approaches either $\pm\infty$, or the point of discontinuity.

Comparison Test. For the comparison test, we have to get very logical. Suppose there is an integral, $\int_a^\infty f(x)dx$, that we cannot integrate and we just want to know whether it converges or diverges. We sometimes can find another integral to compare it to. If we find an integral that has terms always larger than the terms of *our* integral, and that integral **converges,** then *our* integral must converge also. If it did not converge, then at some point it would be larger than the comparison integral which was always bigger and hence, there is a contradiction. Summarizing: If we find an integral that has terms always larger than the terms of the integral we are trying to work with, and it converges, then *our* integral converges also.

If $0 \leq f(x) \leq g(x)$, and $\int_a^\infty g(x)dx$ converges, so does $\int_a^\infty f(x)dx$

Suppose the terms of the integral that we are interested in, $\int_a^\infty g(x)dx$, are always larger than the terms of another integral. If we know that the other integral diverges, then we know that *our* integral diverges also. Otherwise, it couldn't always be greater, right?

If $0 \leq f(x) \leq g(x)$, and $\int_a^\infty f(x)dx$ diverges, so does $\int_a^\infty g(x)dx$

Please note: If we have a divergent integral whose terms are always larger than the terms of the integral we are working with, we can make no conclusion, that is:

$$\text{If } 0 \leq f(x) \leq g(x) \text{ and } \int_a^\infty g(x)dx \text{ diverges,}$$

$$\text{then we can make \textbf{no conclusion} about } \int_a^\infty f(x)dx$$

Also, if we have a convergent integral whose terms are always less than the terms of the integral we are working with, we can make no conclusion.

$$\text{If } 0 \leq f(x) \leq g(x), \text{ and } \int_a^\infty f(x)dx \text{ converges,}$$

$$\text{then we can make \textbf{no conclusion} about } \int_a^\infty g(x)dx$$

Example 1: Examples of Integrals with Discontinuous Integrands

Determine whether or not the following improper integrals are convergent or divergent and evaluate those that are convergent.

$$\text{a)} \quad \int_2^3 \frac{dx}{x-2}$$

$$\text{b)} \quad \int_2^3 \frac{dx}{\sqrt{x-2}}$$

Strategy. In both problems there is a point of discontinuity at $x = 2$. Consequently, we will replace the 2 with the variable a and then evaluate $\lim\limits_{a \to 2^+} \int_a^3 f(x)dx$

Solution. a)

$$\begin{aligned}
\int_2^3 \frac{dx}{x-2} &= \lim_{a \to 2^+} \int_a^3 \frac{dx}{x-2} \\
&= \lim_{a \to 2^+} \ln|x-2| \Big|_a^3 \\
&= \lim_{a \to 2^+} \ln 1 - \ln|a-2| \\
&= 0 - (-\infty) = \infty
\end{aligned}$$

Thus the integral $\int_2^3 \frac{dx}{x-2}$ diverges.

b)

$$\begin{aligned}
\int_2^3 \frac{dx}{\sqrt{x-2}} &= \lim_{a \to 2^+} \int_a^3 \frac{dx}{\sqrt{x-2}} \\
&= \lim_{a \to 2^+} 2\sqrt{x-2} \Big|_a^3 \\
&= \lim_{a \to 2^+} 2\sqrt{1} - 2\sqrt{a-2} \\
&= 2 - 0 = 2
\end{aligned}$$

Thus the integral $\int_2^3 \frac{dx}{\sqrt{x-2}}$ converges to 2.

Summary. When evaluating improper integrals, you always end up finding limits. Often it is helpful to sketch the graph to find the limit. For example, in the first problem, when evaluating $\lim_{a \to 2^+} \ln|a - 2|$, try to picture the graph of the natural logarithm function. If you were having difficulty seeing where the $-\infty$ came from, then look at the graph and you will see the graph approach $-\infty$ as we get closer to the asymptote at $x = 2$.

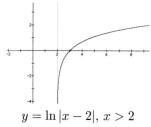

$y = \ln|x - 2|, \; x > 2$

Example 2: Integrals with Infinite Limits

Determine whether or not the following improper integrals are convergent or divergent, and evaluate those that are convergent.

$$\text{a)} \quad \int_0^\infty \frac{dx}{x^2+1}$$

$$\text{b)} \quad \int_{-\infty}^\infty \frac{dx}{x^2+1}$$

Strategy. In these problems we have infinite limits. Since the Fundamental Theorem of Calculus tells us how to evaluate a definite integral with finite limits only, we have to make some changes before we can integrate. We use variables in place of $\pm\infty$ and then take the limit after we evaluate the integrals.

Solution. a)
$$\int_0^\infty \frac{dx}{x^2+1} = \lim_{b\to\infty} \int_0^b \frac{dx}{x^2+1}$$
$$= \lim_{b\to\infty} \tan^{-1} x \Big|_0^b$$
$$= \lim_{b\to\infty} \tan^{-1} b - \tan^{-1} 0$$
$$= \frac{\pi}{2} - 0 = \frac{\pi}{2}$$

Remember $\theta = \tan^{-1} x$ is asking for the angle θ such that $\tan\theta = x$. Consequently, in the above calculation, $\tan^{-1} 0 = 0$ because $\tan 0 = 0$.

b) We first need to divide the integral into two integrals.

$$\int_{-\infty}^\infty \frac{dx}{x^2+1} = \int_{-\infty}^0 \frac{dx}{x^2+1} + \int_0^\infty \frac{dx}{x^2+1}$$

We already know from part a) that $\int_0^\infty \frac{dx}{x^2+1} = \frac{\pi}{2}$.

$$\int_{-\infty}^0 \frac{dx}{x^2+1} = \lim_{a\to-\infty} \int_a^0 \frac{dx}{x^2+1}$$
$$= \lim_{a\to-\infty} \tan^{-1} x \Big|_a^0$$
$$= \lim_{a\to-\infty} (\tan^{-1} 0 - \tan^{-1} a)$$
$$= 0 - \left(-\frac{\pi}{2}\right) = \frac{\pi}{2}$$

Adding the two integrals together, we have

$$\int_{-\infty}^\infty \frac{dx}{x^2+1} = \frac{\pi}{2} + \frac{\pi}{2} = \pi$$

Summary. It helps to picture the graph of the arctangent function when evaluating the limits for $\tan^{-1} x$.

$\lim\limits_{a \to \infty} \tan^{-1} a = \dfrac{\pi}{2}$, since as a approaches $+\infty$, the graph of arctangent approaches $\dfrac{\pi}{2}$.

$\lim\limits_{a \to -\infty} \tan^{-1} a = -\dfrac{\pi}{2}$, since as a approaches $-\infty$, the graph of arctangent approaches $-\dfrac{\pi}{2}$.

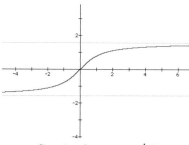

Graph of $x = \tan^{-1} \theta$

Example 3: An Example Using Integration by Parts and L'Hôpital's Rule

Evaluate the improper integral

$$\int_1^\infty \frac{\ln x}{x^2} dx$$

Strategy. Obviously, this is an improper integral since the upper limit is ∞. We use the variable b as the upper limit, evaluate the integral and then take the limit as $b \to \infty$. When we take this limit, we need to use L'Hôpital's Rule.

Solution.

$$\int_1^\infty \frac{\ln x}{x^2} dx = \lim_{b \to \infty} \int_1^b \frac{\ln x}{x^2} dx$$

We use integration by parts with

$$u = \ln x, \quad dv = \frac{1}{x^2} dx$$
$$du = \frac{dx}{x}, \quad v = -\frac{1}{x}$$

$$\int_1^b \frac{\ln x}{x^2} dx = \underbrace{-\frac{1}{x} \ln x}_{uv} \Big|_1^b - \int_1^b \underbrace{\left(-\frac{1}{x}\right)}_{v} \underbrace{\left(\frac{1}{x}\right) dx}_{du}$$

$$= -\frac{\ln b}{b} - 0 + \int_1^b \frac{1}{x^2} dx$$

$$= -\frac{\ln b}{b} + -\frac{1}{x} \Big|_1^b$$

$$= -\frac{\ln b}{b} + \left(-\frac{1}{b}\right) - (-1)$$

$$= -\frac{\ln b}{b} - \frac{1}{b} + 1$$

$$\Longrightarrow \int_1^\infty \frac{\ln x}{x^2} dx = \lim_{b \to \infty} \left(-\frac{\ln b}{b} - \frac{1}{b} + 1\right)$$

Since $\lim_{b \to \infty} \frac{\ln b}{b}$ is of type $\frac{\infty}{\infty}$, we use L'Hôpital's Rule to find the limit.

$$\lim_{b \to \infty} \frac{\ln b}{b} = \lim_{b \to \infty} \frac{\frac{1}{b}}{1} = 0$$

It now follows that

$$\int_1^\infty \frac{\ln x}{x^2} dx = \lim_{b \to \infty} \left(-\frac{\ln b}{b} - \frac{1}{b} + 1\right) = -0 - 0 + 1 = 1$$

Summary. There is a lot going on in this problem. It is important to keep track of where you are and what you are doing so that you can put it all back together again like a puzzle.

Example 4: Using the Comparison Test to Show that an Integral Converges

Use the comparison test to show that the following integral converges.

$$\int_1^\infty \frac{dx}{x^4+1}$$

Strategy. In order to show that the integral converges, we must find a function that is always larger than $f(x) = \frac{1}{x^4+1}$, and whose integral converges. Remember that in a fraction, if the denominator gets larger, the fraction gets smaller.

Solution. First of all we have no method to integrate the integral $\int \frac{dx}{x^4+1}$. However, the integrand looks similar to $\frac{1}{x^4}$. For all x we know that

$$x^4 + 1 > x^4 \implies \frac{1}{x^4+1} < \frac{1}{x^4}$$

Therefore, if we can show that $\int \frac{1}{x^4} dx$ converges, then we can conclude that $\int \frac{1}{x^4+1} dx$ also converges. Let's see whether or not the integral $\int_1^\infty \frac{dx}{x^4}$ converges.

$$\begin{aligned}
\int_1^\infty \frac{dx}{x^4} &= \lim_{b \to \infty} \int_1^b \frac{dx}{x^4} \\
&= \lim_{b \to \infty} \left. -\frac{1}{3} x^{-3} \right|_1^b \\
&= \lim_{b \to \infty} \left(-\frac{1}{3b^3} + \frac{1}{3} \right) = 0 + \frac{1}{3} = \frac{1}{3}
\end{aligned}$$

Thus $\int_1^\infty \frac{dx}{x^4}$ converges, and by the comparison test so does $\int_1^\infty \frac{dx}{x^4+1}$.

Summary. When we have a rational integrand, it is usually helpful to look for an integral that we know converges with a smaller denominator than the one we have. Then the given fraction will always be less than the one we are going to compare it to.

Example 5: Using the Comparison Test to Show that an Integral Diverges

Use the comparison theorem to show that the following integral diverges.
$$\int_1^\infty \frac{dx}{x - \sqrt{x}}$$

Strategy. This is similar to the previous example in that we need an integral to compare this one to. However, in this example, we have to find a function that is smaller than $\frac{1}{x-\sqrt{x}}$, and whose integral diverges. One function to compare $\frac{1}{x-\sqrt{x}}$ to is $y = \frac{1}{x}$. We get for all $x \geq 1$,

$$x > x - \sqrt{x} \implies \frac{1}{x} < \frac{1}{x - \sqrt{x}}$$

If we can show that the smaller integral $\int_1^\infty \frac{dx}{x}$ diverges, then we know that *our* integral also diverges.

Solution. Here we will compare the integral to
$$\int_1^\infty \frac{dx}{x} = \lim_{b \to \infty} \int_1^b \frac{dx}{x} = \lim_{b \to \infty} (\ln b - \ln 1) = \infty - 0 = \infty$$

Since $\frac{1}{x-\sqrt{x}} > \frac{1}{x} \geq 0$, it follows by comparison that since $\int_1^\infty \frac{dx}{x}$ diverges, so does $\int_1^\infty \frac{dx}{x-\sqrt{x}}$.

Summary. Remember that if you want to show that an integral diverges, then you must find an integral which diverges where the integrand is always smaller than the integrand that you are working with. You can picture it as though the smaller integral is diverging, meaning that it is going off to infinity and *pushing* the larger integral to infinity with it.

CHAPTER 5

Applications of the Definite Integral

5.1. Area of a Plane Region

Key Topics.
- Area Between two Curves

Worked Examples.
- Finding the Area Between two Curves
- An Area or a Region Bounded by Functions of y

Overview

When we first presented integrals we talked about the area that was represented by the region between a curve and the x-axis. If we remember that the x-axis represents the curve $y = 0$, then we were actually finding the area of the region between the given function and $y = 0$.

Area Between two Curves. It is not a major leap to go from the area between a curve and the x-axis to the area between two curves. We will sketch a sample rectangle with height the distance between the two curves.

If the two functions are defined in terms of the variable x, then we draw a *vertical* rectangle. The width, change in x, is denoted by Δx. The height is **always** found by subtracting top function minus bottom function. The area is then

$$(f_{\text{top}}(x) - f_{\text{bottom}}(x)) \cdot \Delta x$$

If we add up all the possible sample rectangles which can be formed over the interval $x = a$ to $x = b$, then we will get the area of the region as:

$$A = \int_a^b (f_{\text{top}}(x) - f_{\text{bottom}}(x))\, dx$$

If the functions are defined in terms of y, then we draw a *horizontal* rectangle. The width, change in y, is denoted by Δy. The height, or length, is found by subtracting right function minus left function. The area of the sample rectangle is then

$(f_{\text{right}}(y) - f_{\text{left}}(y)) \cdot \Delta y$. If we add up all the sample rectangles which can be formed over the interval $y = c$ to $y = d$, then we will get the area of the region as:

$$A = \int_c^d \left(f_{\text{right}}(y) - f_{\text{left}}(y)\right) dy$$

Be sure and note that if we use dx, then all the terms must be in terms of x and the limits of integration are x-limits.

Similarly, if we use dy, all the terms must by in terms of y and the limits of integration are y-limits.

Example 1: Finding the Area Between two Curves

Find the area of the region enclosed by the curves $y = x^2$ and $y = 2 - x^2$.

Strategy. Since both curves are given in terms of x, we draw our sample rectangle vertically. Consequently, the width is Δx and we proceed making sure we have the functions in terms of x and the limits as x-limits.

Solution. First we sketch the region enclosed by the two parabolas $y = x^2$ and $y = 2 - x^2$. We slice this region into vertical rectangular slices.

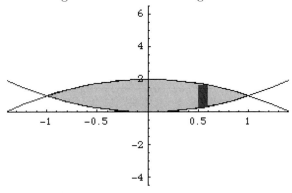

As stated above, the width of a typical rectangular slice is Δx. To obtain the height of the rectangle, we must subtract the bottom curve ($y = x^2$) from the top curve ($y = 2 - x^2$). Remember that we always find the height of a vertical rectangle by subtracting top minus bottom.

$$\text{height} = (2 - x^2) - x^2 = 2 - 2x^2$$
$$\text{width} = \Delta x$$

Now we need to find the limits of integration. We need to identify the x-coordinates of the points where the two curves intersect. Consequently, we solve the equation:

$$x^2 = 2 - x^2$$
$$2x^2 = 2$$
$$x^2 = 1$$
$$x = \pm 1$$

We are slicing the area into rectangles from $x = -1$ to $x = 1$ making the limits of integration -1 and $+1$.

$$\begin{aligned}\text{Area} &= \int_{-1}^{1} \left(2 - 2x^2\right) dx \\ &= \left. 2x - \frac{2}{3}x^3 \right|_{-1}^{1} \end{aligned}$$

$$= \left(2 - \frac{2}{3}\right) - \left(-2 + \frac{2}{3}\right) = \frac{4}{3} + \frac{4}{3} = \frac{8}{3}$$

Summary. It really is important to sketch the region and a sample rectangle. Obviously, there cannot be a negative area. If you integrate and end up with a negative number, go back and make sure that you did not subtract *backwards*.

Example 2: An Area or a Region Bounded by Functions of y

Find the area of the region enclosed by the curves $x = 6 - y^2$ and $x = y$.

Strategy. In this problem, the first function is written in terms of y. If we solve for y in terms of x, we get $y = \pm\sqrt{6-x}$. The $\pm$ in front of the radical, leads us to think that we are better off keeping the function in terms of y, and drawing *horizontal* rectangles. Note that this problem can be done either way, but it will be easier to solve it in terms of y.

Solution. First we sketch the region.

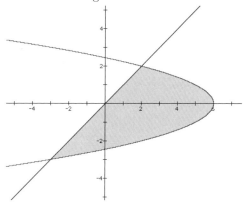

We slice the region horizontally so that the width is Δy, and we can set up an integral for the area using the y-variable.

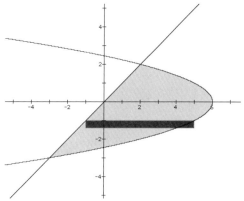

To get the height (or length) of this rectangle we must subtract the function on the left ($x = y$) from the function on the right ($x = 6 - y^2$). Remember that we always find the height of a horizontal rectangle by subtracting right minus left.

$$\begin{aligned} \text{height} &= 6 - y^2 - y = -y^2 - y + 6 \\ \text{width} &= \Delta y \end{aligned}$$

Now we need to find the limits of integration. We need to identify the y-coordinate of the points where the two curves intersect. Consequently, we solve the equation:
$$\begin{aligned} y &= 6 - y^2 \\ y^2 + y - 6 &= 0 \\ (y+3)(y-2) &= 0 \\ \implies y &= -3, \text{ and } y = 2 \end{aligned}$$

Thus, we are slicing the area into horizontal rectangles from $y = -3$ to $y = 2$ and the limits of integration are -3 and 2.

$$\begin{aligned} \text{Area} &= \int_{-3}^{2} (-y^2 - y + 6) \, dy \\ &= \left. -\frac{1}{3} y^3 - \frac{1}{2} y^2 + 6y \right|_{-3}^{2} \\ &= \left(\frac{-1}{3}(8) - \frac{4}{2} + 12 \right) - \left(\frac{27}{3} - \frac{9}{2} - 18 \right) \\ &= -\frac{8}{3} - 2 + 12 - 9 + \frac{9}{2} + 18 \\ &= \frac{125}{6} \end{aligned}$$

Summary. Once again it is important to sketch the region and a sample rectangle. Make sure that you subtract right function minus left function and make sure you use the y-limits.

In the next section, we will see how we can use integrals to find the volume of a solid.

5.2. Volume

Key Topics.
- Volumes by Disks
- Volumes by Washers
- Volumes by Cylindrical Shells

Worked Examples.
- Using Disks and Washers to Find the Volume of the Solid of Revolution for a Region bounded by $y = f(x)$
- Using Disks and Washers to Find the Volume of the Solid of Revolution for a Region bounded by $x = f(y)$
- Using Shells to find the Volume

Overview

In this section we use integrals in order to find the volume of solids. The text has some nice examples on computing volumes from cross-sectional areas. We will concentrate here on finding the volume of solids of revolution. A solid of revolution is the solid formed by rotating, or revolving, a plane region about a line called the axis of revolution, or the axis of rotation. We will rotate about the x-axis, the y-axis, and lines of the form $x = a$ or $y = b$ where a and b are constants. It is imperative that you make sketches in order to determine what the regions look like and which method will work best.

Volumes by Disks. If one boundary of the region lies on the axis of rotation, then we often use the disk method. Draw a sample rectangle **perpendicular** to the axis of rotation. If you picture this rectangle rotating about the axis, you realize that it makes a circle. Since the area of a circle is $A = \pi r^2$, all we need to have to find the volume is the thickness. If there is a vertical rectangle, then the thickness is dx and if there is a horizontal rectangle, then the thickness is dy. Once we know whether we need dx or dy then we know which variable to use throughout the problem. If you really can picture the circle, then it will be easier to remember the formula for volume using disks: $V = \pi r^2 \cdot$ (thickness), where r, the radius, is the distance from one end of the rectangle to the other end of the rectangle which is either $f(x) - 0$, or $g(y) - 0$.

$$V = \int_a^b \pi (f(x))^2 dx \quad \text{or} \quad V = \int_c^d \pi (g(y))^2 dy$$

Volumes by Washers. Solids with a hole in the middle (picture a doughnut) are formed when the region does not have one side on the axis of revolution. As a result, when the region rotates, there is a hole in it. Once again, the rectangle is drawn **perpendicular** to the axis of rotation. We need to compute the volume formed using the outer radius (distance from the axis of rotation to the furthest part of the rectangle), and then subtract the volume formed using the inner radius (distance from the axis of rotation to the closest part of the rectangle). In other words we need:

$$(\pi r_{\text{outer}}^2 - \pi r_{\text{inner}}^2) \cdot \text{thickness} = \pi(r_{\text{outer}}^2 - r_{\text{inner}}^2) \cdot \text{thickness}$$

Note: $\pi(r_{\text{outer}}^2 - r_{\text{inner}}^2)$ is **not** the same as $\pi(r_{\text{outer}} - r_{\text{inner}})^2$

$$V = \int_a^b \pi(r_{\text{outer}}^2 - r_{\text{inner}}^2)dx \quad \text{or} \quad V = \int_c^d \pi(r_{\text{outer}}^2 - r_{\text{inner}}^2)dy$$

Volumes by Cylindrical Shells. If you think of a roll of paper towels, you will know what volumes by shells look like. Note that the paper towels are **parallel** to the cardboard tube which is the axis of revolution. To use volume by shells, draw the sample rectangle **parallel** to the axis of revolution. If you picture that rectangle rotating around the axis, it forms a cylindrical shell. If you *open up* this shell it becomes a rectangle with its length the circumference of the circular base which is $2\pi r$, the height of the rectangle and the thickness become either dx or dy. (Picture unrolling the paper towels and looking at one sheet.). So to obtain the volume by cylindrical shells, we need either:

$$V = \int_a^b 2\pi rh\, dx \quad \text{or} \quad V = \int_c^d 2\pi rh\, dy$$

where the radius is the distance from the axis of rotation to the sample rectangle.

Use the diagrams below to decide which method to use.

```
        s h e l l
     p a r a l l e l


                                    w
                                    a
                                    s
                                    h
                                    e
      p e r p e n d i c u l a r
                        i
                        s
                        k
```

Example 1: Using Disks and Washers to Find the Volume of the Solid of Revolution for a Region bounded by $y = f(x)$

Find the volume of the solid generated by revolving the region bounded by the parabola $y = 4 - x^2$, the x-axis and the y-axis about a) the x-axis and b) the line $y = -1$.

Strategy. We sketch the region and note that the function is given in terms of the variable x. We slice the region into vertical slices which are perpendicular to both of the axes of rotation which are horizontal lines. Therefore, we know that we need to use either disks or washers for both problems.

Solution. a) First we sketch the region, and a typical slice of the solid of revolution.

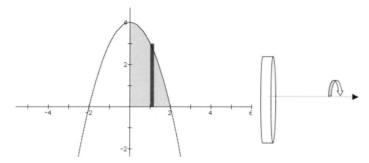

In this part, we are rotating about the x-axis. The region lies on the x-axis, thus there is no hole in the solid, and hence we are using disks. We need

$$\text{radius} = 4 - x^2$$
$$\text{thickness} = dx$$

We are slicing the region along the x-axis from $x = 0$ to $x = 2$. The volume is given by:

$$\int_0^2 \pi \left(4 - x^2\right)^2 dx$$
$$= \int_0^2 (16 - 8x^2 + x^4) dx$$
$$= \pi \left[16x - \frac{8}{3}x^3 + \frac{1}{5}x^5\right]\Big|_0^2$$
$$= \pi \left[32 - \frac{64}{3} + \frac{32}{5} - 0\right]$$
$$= \frac{256}{15}\pi$$

b) Again, we start this problem with a picture.

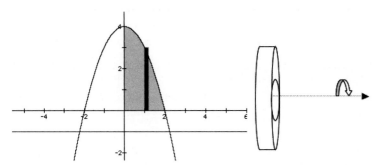

When rotating this region about the line $y = -1$, there will be a hole in it. Hence, we use the washer method. We still draw the rectangle **perpendicular** to the axis of rotation and we have

$$\text{outer radius} = r_{\text{outer}} = 4 - x^2 - (-1) = 5 - x^2$$
$$\text{inner radius} = r_{\text{inner}} = 0 - (-1) = 1$$
$$\text{thickness} = dx$$

$$\begin{aligned}
V &= \int_0^2 \pi \left[(5 - x^2)^2 - (1)^2\right] dx \\
&= \pi \int_0^2 \left(25 - 10x^2 + x^4 - 1\right) dx \\
&= \pi \int (24 - 10x^2 + x^4) dx \\
&= \pi \left(24x - \frac{10}{3}x^3 + \frac{1}{5}x^5\right)\bigg|_0^2 \\
&= \pi \left(48 - \frac{80}{3} + \frac{32}{5} - 0\right) \\
&= \frac{416}{15}\pi
\end{aligned}$$

Summary. A sketch is mandatory for the volume problems. Also, as stressed in your text, do NOT attempt to memorize what to do. You will end up with a lot of formulas with no meanings attached to them, and will not know when to use what formula. If you study the geometry behind the formulas, then these problems are quite logical.

Example 2: Using Disks and Washers to Find the Volume of the Solid of Revolution for a Region bounded by $x = f(y)$

Find the volume of the solid generated by revolving the region bounded by $x = \sqrt{y}$, and $x = \dfrac{y}{3}$ about a) the y-axis and b) the line $x = 10$.

Strategy. We are going to slice the region into horizontal slices since the functions are given in term of the variable, y. We are rotating about vertical lines, which means the rectangles are perpendicular to the axes of rotation. Therefore, we need to use either disks or washers.

Solution. a) We first sketch the region.

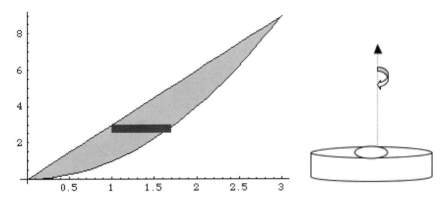

a) When we rotate this region about the y-axis, there will be a hole. Thus, we are using washers.

$$\begin{aligned} \text{outer radius} &= r_{\text{outer}} = \sqrt{y} \\ \text{inner radius} &= r_{\text{inner}} = \frac{y}{3} \\ \text{thickness} &= dy \end{aligned}$$

Since the curves intersect at $(0,0)$ and $(3,9)$, and since we are using y-limits, we get the limits of integration are $y = 0$, and $y = 9$. Thus

$$\begin{aligned} V &= \int_0^9 \pi \left((\sqrt{y})^2 - \left(\frac{y}{3}\right)^2 \right) dy \\ &= \pi \int_0^9 \left(y - \frac{1}{9} y^2 \right) dy \\ &= \pi \left(\frac{1}{2} y^2 - \frac{1}{27} y^3 \right) \Big|_0^9 \\ &= \pi \left(\frac{81}{2} - 27 - 0 \right) \\ &= \frac{27}{2} \pi \end{aligned}$$

b) The line $x = 10$ is also a vertical line. The rectangles are horizontal once again, so that we use the variable, y. Using washers, we get

$$\begin{aligned} \text{outer radius} &= r_{\text{outer}} = 10 - \frac{y}{3} \\ \text{inner radius} &= r_{\text{inner}} = 10 - \sqrt{y} \\ \text{thickness} &= dy \end{aligned}$$

Thus the volume is equal to

$$\begin{aligned} V &= \pi \int_0^9 \left[\left(10 - \frac{y}{3}\right)^2 - (10 - \sqrt{y})^2 \right] dy \\ &= \pi \int_0^9 \left(100 - \frac{20}{3}y + \frac{y^2}{9} - 100 + 20\sqrt{y} - y \right) dy \\ &= \pi \int_0^9 \left(\frac{1}{9}y^2 - \frac{23}{3}y + 20y^{\frac{1}{2}} \right) dy \\ &= \pi \left(\frac{1}{27}y^3 - \frac{23}{6}y^2 + \frac{40}{3}y^{\frac{3}{2}} \right) \Big|_0^9 \\ &= \pi \left(\left(27 - \frac{621}{2} + 360 \right) - (0 - 0 + 0) \right) \\ &= \frac{153}{2}\pi \\ &= 76.5\pi \end{aligned}$$

Summary. Remember that once you have horizontal rectangles and dy, the integrand and the limits of integration must all be in terms of y.

Example 3: Using Shells to find the Volume

Set up the integral for the volume of the solid generated by revolving the region bounded by $y = 8x - 2x^2$ about the y-axis.

Strategy. First sketch the region. The function is given in terms of the variable x, so we need a vertical slice. We are rotating about the y-axis. Since we have a vertical rectangle which is parallel to the axis of rotation, we need to use shells.

Solution.

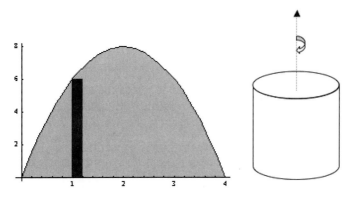

We use the shell method,

$$\begin{aligned} \text{radius} &= x \\ \text{height} &= (8x - 2x^2) - 0 = 8x - 2x^2 \\ \text{thickness} &= dx \end{aligned}$$

Notice that the radius, which is the distance from the y-axis to the sample rectangle, is simply the horizontal distance from the y-axis, which is x. The volume is:

$$V = \int_0^4 2\pi \underbrace{\left(x\right)}_{\text{radius}} \underbrace{\left(8x - 2x^2\right)}_{\text{height}} dx$$

Summary. Make sure you understand why we had to use the shell method here. It would have been difficult to rewrite the function in terms of y, so the rectangle had to have width dx. This rectangle is parallel to the y-axis and hence makes cylindrical shells.

5.3. Arc Length and Surface Area

Key Topics.
- Arc Length
- Surface Area

Worked Examples.
- Arc Length of a Function
- Arc Length of a Parametric Function
- Surface Area

Overview

We have found area and volume, but there are two other measurements that we still need to examine. Here we will look at the length of an arc of a curve, and the area of the surface of a solid.

Arc Length. When we talk about the length of an arc, picture making the curve out of a piece of string. If you take the piece of string and straighten it out, then the length could easily be measured. Here we use the formula for arc length, s.

$$s = \int_a^b \sqrt{1 + [f'(x)]^2}\, dx$$

Surface Area. When we found the volume of the solids of revolution in the last section, we rotated a region about an axis. The area of a surface is found by rotating an arc of a curve about an axis. It is similar to the area formed by the rope when a person is jumping rope. If we partition the arc into lots of little teeny intervals, they are each like little dots rotating around the axis and forming the circumference of a circle. If the radius is $f(x)$, then each circumference is $2\pi f(x)$. Since we need the entire length of the arc on the interval $[a, b]$, the formula for surface area is:

$$S = \int_a^b 2\pi f(x) \cdot \sqrt{1 + [f'(x)]^2}\, dx$$

Example 1: Arc Length of a Function

Find the length of the curve $y = 2(x-2)^{\frac{3}{2}}$, $2 \leq x \leq 9$.

Strategy. We use the formula for arc length, and integrate using substitution.

Solution. $f(x) = 2(x-1)^{\frac{3}{2}}$, thus $f'(x) = 2 \cdot \frac{3}{2}(x-2)^{\frac{1}{2}} = 3(x-2)^{\frac{1}{2}}$

$$\begin{aligned}
\text{Arc Length} &= \int_2^9 \sqrt{1 + [f'(x)]^2} \, dx \\
&= \int_2^9 \sqrt{1 + \left[3(x-2)^{\frac{1}{2}}\right]^2} \, dx \\
&= \int_2^9 \sqrt{1 + 9(x-2)} \, dx \\
&= \int_2^9 \sqrt{9x - 17} \, dx
\end{aligned}$$

We can use a substitution,

$$\begin{aligned}
\text{if } u &= 9x - 17 \\
\text{then } du &= 9dx \\
\text{and } \frac{1}{9} du &= dx
\end{aligned}$$

$$\begin{aligned}
& \frac{1}{9} \int_{u=1}^{u=64} u^{\frac{1}{2}} \, du \\
&= \frac{1}{9} \cdot \frac{2}{3} u^{\frac{3}{2}} \Big|_1^{64} \\
&= \frac{2}{27} \left(64^{\frac{3}{2}} - 1^{\frac{3}{2}}\right) \\
&= \frac{2}{27} \left(\left(\sqrt{64}\right)^3 - 1\right) \\
&= \frac{2}{27} (512 - 1) = \frac{1022}{27}
\end{aligned}$$

Summary. We were able to easily integrate this integral for arc length. Many of the integrals that we obtain by using the formula for arc length are very difficult or impossible to integrate by any method other than numerical integration. In those cases use a computer or calculator to evaluate those integrals.

Example 2: Arc Length of a Parametric Function

Find the length of the parametric curve given by
$$\begin{cases} x = 3\cos t \\ y = 3\sin t \end{cases} \text{ for } 0 \leq t \leq \pi$$

Strategy. Here we have a parametric curve, where both x and y are defined in terms of t, which often stands for time. We use the formula for arc length of a parametric function.
$$S = \int_a^b \sqrt{[x'(t)]^2 + [y'(t)]^2}\, dt$$

Solution. First we must find the two derivatives, $x'(t) = \dfrac{dx}{dt}$ and $y'(t) = \dfrac{dy}{dt}$

$$\frac{dx}{dt} = -3\sin t \text{ and } \frac{dy}{dt} = 3\cos t$$

$$\begin{aligned}
\text{Arc Length} &= \int_0^\pi \sqrt{(-3\sin t)^2 + (3\cos t)^2}\, dt \\
&= \int_0^\pi \sqrt{9\sin^2 t + 9\cos^2 t}\, dt \\
&= \int_0^\pi 3\, dt = 3\big|_0^\pi = 3\pi
\end{aligned}$$

Summary. Once again we have an integral that is easily integrated. Following the formula, hopefully, is not very difficult.

Example 3: Surface Area

Find the surface area generated by revolving $y = \sqrt{x}$ for $0 \leq x \leq 2$ about the x-axis.

Strategy. We use the formula for surface area:
$$S = \int_a^b 2\pi f(x) \sqrt{1 + [f'(x)]^2}\,dx$$
Then we will find a common denominator for the expression under the radical sign and simplify algebraically. At that point, we will substitute and integrate.

Solution. We need:
$$\begin{aligned} f(x) &= \sqrt{x} = x^{\frac{1}{2}} \\ f'(x) &= \frac{1}{2} x^{\frac{-1}{2}} \\ [f'(x)]^2 &= \frac{1}{4} x^{-1} = \frac{1}{4x} \end{aligned}$$

Now putting all the parts into the formula we have:
$$\begin{aligned} S &= \int_0^2 2\pi\sqrt{x}\sqrt{1 + \frac{1}{4x}}\,dx \\ &= \int_0^2 2\pi\sqrt{x}\sqrt{\frac{4x+1}{4x}}\,dx \\ &= \int_0^2 2\pi\sqrt{x}\frac{\sqrt{4x+1}}{2\sqrt{x}}\,dx \\ &= \pi \int \sqrt{4x+1}\,dx \end{aligned}$$

Now substitute
$$\begin{aligned} \text{if } u &= 4x+1 \\ \text{then } du &= 4dx \\ \text{and } \frac{1}{4}du &= dx \end{aligned}$$

Also, the limits of integration will change to $u = 1$ to $u = 9$.
$$\begin{aligned} \frac{\pi}{4}\int_{u=1}^{u=9} u^{\frac{1}{2}}\,du &= \frac{\pi}{4}\cdot\frac{2}{3}u^{\frac{3}{2}}\Big|_1^9 \\ &= \frac{\pi}{6}\cdot(9^{\frac{3}{2}} - 1) \\ &= \frac{\pi}{6}\cdot(\sqrt{9}^3 - 1) \\ &= \frac{13}{3}\pi \end{aligned}$$

Summary. Try to make sense of the meaning of these formulas and how they are derived. In this way, it will be much easier to remember them.

5.4. Projectile Motion

Key Topics.
- Vertical Motion
- Projectile Motion

Worked Examples.
- Vertical Motion of a Chocolate Pie
- Projectile Motion of a Cannonball

Overview

In this section we will see how to use integrals to help solve problems dealing with projectile motion. Projectile motion deals with the position, velocity and acceleration of an object that has been set in motion by something like jumping, shooting or throwing.

Vertical Motion. When we were learning derivatives, we found that if $h(t)$ is the position function standing for height above the ground, then the velocity function, $v(t)$ is simply the change in position or $h'(t)$. Similarly, the acceleration is the change in velocity, and hence $a(t) = v'(t) = h''(t)$.

One of the great facts about integration is that if we start with acceleration, we can use integration to work our way back to the position.

We use acceleration due to gravity as $-9.8 \, \text{m/s}^2$ or $-32 \, \text{ft/s}^2$ depending upon which system of measurement we are using.

To summarize, if we start with $a(t)$, then $v(t) = \int a(t)dt$ and $h(t) = \int v(t)dt$.

Projectile Motion. If our projectile travels not just vertically, but horizontally also, then we separate our functions into two components - the horizontal component and the vertical component. We denote these by using subscripts.

If $y(t)$ and $x(t)$ represent the vertical and horizontal positions respectively, then $y''(t) = -g$ where g is the gravitational constant as before. In this case $x''(t) = 0$ since gravity does not have a horizontal pull. To summarize:

$$a_x(t) = x''(t) = 0$$
$$a_y(t) = y''(t) = -g$$

If the projectile is fired at an angle of $\theta \geq 0$, then we can separate the velocity into two components using right triangle trigonometry.

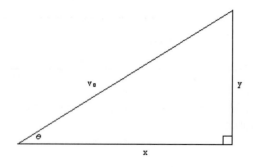

$$\sin\theta = \frac{y}{v_0} \quad \implies \quad y = v_0 \sin\theta$$
$$\cos\theta = \frac{x}{v_0} \quad \implies \quad x = v_0 \sin\theta$$

Example 1: Vertical Motion of a Chocolate Pie

A chocolate pie is thrown directly upward at a speed of 64 feet per second from a platform 80 feet above the ground. At what time does the pie reach its highest point? How high does it go? When does the pie hit the ground? What is the velocity at the instant the pie hits the ground?

Strategy. Since the pie is thrown directly upward, we only have to deal with the vertical change. We start with the initial conditions and work from there.

Solution. The initial information given can be summarized as follows:

$$\begin{cases} a = -32 \\ v_0 = 64 \\ h_0 = 80 \end{cases}$$

To find the velocity function, we integrate $a(t)$. $v(t) = \int -32 dt = -32t + c$, and since $v(0) = 64$, we have $c = 64$, thus

$$v(t) = -32t + 64$$

To find the position function $h(t)$, we integrate $v(t)$. $h(t) = \int(-32t + 64)dt = -16t^2 + 64t + c$, and since $h(0) = 80$, we have $c = 80$, therefore

$$h(t) = -16t^2 + 64t + 80$$

Now that we have all three functions, we can proceed to answer the questions:

a) At what time does the pie reach its highest point?
This asks us when $h(t)$ is a maximum. Consequently, we must solve the equation: $h'(t) = v(t) = 0$

$$v(t) = -32t + 64 = 0 \implies t = 2\,\text{s}$$

Therefore, the pie reaches its highest point at time $t = 2$ seconds.

b) How high does it go?
We know that it reaches the highest point at $t = 2\,\text{s}$. Now we need to find its position at this time.

$$h(2) = -16(2)^2 + 64(2) + 80 = 144\,\text{ft}$$

c) When does the pie hit the ground?
The pie is on the ground when $h(t)$, which represents the height above the ground, is equal to zero.

$$\begin{aligned} -16t^2 + 64t + 80 &= 0 \\ -16(t^2 - 4t - 5) &= 0 \\ -16(t - 5)(t + 1) &= 0 \end{aligned}$$

$t = 5\,\text{s}$ is the only answer which makes sense in this problem. The pie hits the ground after $t = 5$ seconds.

d) What is the velocity at the instant the pie hits the ground?
Since the pie hits the ground at time $t = 5\,\text{s}$, we are looking for $v(5)$.
$$v(5) = -32(5) + 64 = -96\,\text{ft/s}$$
The velocity at impact is $-96\,\text{ft/s}$.

Summary. The fact that the velocity is negative tells us that the pie is moving in a downward motion, which is true.

Once we have found all three functions, then we just need to use thought and logic to determine which function to use to answer the given questions.

Example 2: Projectile Motion of a Cannonball

A cannon is fired with an angle of elevation of 45 degrees. The speed of the cannonball as it leaves the muzzle is 128 ft/s. What is the maximum height of the cannonball? How long will the cannonball be in the air? How far away from the muzzle is the cannonball when it lands?

Strategy. Notice how this example differs from the previous one. The cannonball is moving in both a vertical and a horizontal direction. We must, therefore, work with the horizontal and vertical components separately.

Solution. Since the cannonball starts at ground level, the initial position is
$$\begin{cases} h_x(0) = 0 \\ h_y(0) = 0 \end{cases}$$

Also, since it has not been shot yet, the initial horizontal acceleration is 0, while the vertical acceleration due to gravity is $-32 \, \text{ft}/\text{s}^2$.

This information can be summarized as follows:
$$\begin{cases} a_x = 0 \\ v_x(0) = 128 \cos(\frac{\pi}{4}) = 128 \cdot \frac{\sqrt{2}}{2} = 64\sqrt{2} \\ h_x(0) = 0 \end{cases}$$

and
$$\begin{cases} a_y = -32 \\ v_y(0) = 128 \sin(\frac{\pi}{4}) = 128 \cdot \frac{\sqrt{2}}{2} = 64\sqrt{2} \\ h_y(0) = 0 \end{cases}$$

First we need to find all the functions we will use.
At the beginning, $a_x = 0$ and $v_x = 64\sqrt{2}$. We find h_x by integrating v_x.
$$h_x = \int 64\sqrt{2} \, dt = 64\sqrt{2}t + c$$
Since $h(0) = 0$, we have $c = 0$ and
$$h_x = 64\sqrt{2}t$$

Similarly, at the beginning, $a_y = -32$. We find v_y by integrating a_y.
$$v_y = \int -32 \, dt = -32t + c$$
Since $v_y(0) = 64\sqrt{2}$, we have $c = 64\sqrt{2}$ and
$$v_y = -32t + 64\sqrt{2}$$

We can find h_y by integrating v_y.
$$h_y = \int (-32t + 64\sqrt{2}) \, dt = -16t^2 + 64\sqrt{2}t + c$$

Since $h_y(0) = 0$, we have
$$h_y = -32t + 64\sqrt{2}$$
Putting these all together, we have:
$$\begin{cases} a_x = 0 \\ v_x = 64\sqrt{2} \\ h_x = 64\sqrt{2}t \end{cases} \text{and} \begin{cases} a_y = -32 \\ v_y = -32t + 64\sqrt{2} \\ h_y = -16t^2 + 64\sqrt{2}t \end{cases}$$

Now we can answer the questions.

a) What is the maximum height?

The maximum height occurs when $h_y'(t) = v_y(t) = 0$.
$$\begin{aligned} v_y(t) &= -32t + 64\sqrt{2} = 0 \\ -32t &= -64\sqrt{2} \\ t &= 2\sqrt{2}\,\text{s} \end{aligned}$$

The actual height at this time is:
$$\begin{aligned} h_y(2\sqrt{2}) &= -16(2\sqrt{2})^2 + 64\sqrt{2}(2\sqrt{2}) \\ &= -16(8) + 64(4) = 128\,\text{ft} \end{aligned}$$

b) How long is the cannonball in the air?

If we find the time when it hits the ground, then we will know how long it has been in the air since it starts on the ground and ends on the ground.

We need to solve the equation $h_y = 0$.
$$\begin{aligned} -16t^2 + 64\sqrt{2}t &= 0 \\ -16t(t - 4\sqrt{2}) &= 0 \\ t &= 0 \text{ and } t = 4\sqrt{2} \end{aligned}$$

From $t = 0$ we know that the cannonball began at ground level.

From $t = 4\sqrt{2}$ we know that it was in the air for $4\sqrt{2}$ s.(before it returned to the ground).

c) How far away from the muzzle is the cannonball when it lands?

This question refers to the horizontal distance the cannonball travels. Since we know that it took $4\sqrt{2}$ seconds to land, we need to evaluate $h_x(4\sqrt{2})$.
$$\begin{aligned} h_x(4\sqrt{2}) &= 64\sqrt{2}(4\sqrt{2}) \\ &= 64(8) = 512\,\text{ft} \end{aligned}$$

The cannonball landed 512 feet from the muzzle.

Summary. Once again we find all the functions we need, then use thought and logic to determine which function will give the answer to the question we are asked.

5.5. Applications of Integration to Physics and Engineering

Key Topics.
- Work
- Center of Mass

Worked Examples.
- A Spring Example
- A Leaky Bucket
- Filling a Tank
- Mass and Center of Mass

Overview

In this section we apply the technique of integration to physics and engineering problems. Note that in all cases we partition an interval, find a Riemann sum, and then take a limit and obtain an integral. This is the process which defines integration.

Work. Mathematically, work is defined as force times distance. It is not the word *work* as we are used to using in everyday language. According to this definition, studying or teaching is not work. Work is only the movement of some weight over some distance. If the force is variable, then we need to partition an interval $[a,b]$, find the work at each little part, take the limit as the number of partitions approaches infinity, and then add it all together, which, in other words, is integration.

$$W = \int_a^b F(x)dx$$

Center of Mass. The center of mass is basically the *balance point*. In a simplified example, on a seesaw, if there are two children of equal weight sitting the same distance from the center on opposite sides of the center, then the center of mass is the pivot point equidistant from both children.

The point, which is the center of mass, acts as if all the mass is concentrated in that point.

Example 1: A Spring Example

A force of 9 ft-lb (foot-pounds) is required to hold a spring that has been stretched from its natural length of 2 in to a length of 5 in. How much work is done in stretching the spring from 6 in to 12 in?

Strategy. Hooke's Law states that the force, F, required to compress or stretch a spring is proportional to the distance, d, that the spring is compressed or stretched from its original length. In symbolic terms, $F = kd$, where k is a constant of proportionality dependent on the specific spring.

Solution. This problem mixes feet and inches. We must use consistent units. We will use feet as the measure of length with the conversion factor 12 in = 1 ft. We are told that a force of 9 ft-lb is needed to stretch a spring from 2 in to 5 in. This represents a stretch of 3 in or $\frac{3}{12} = \frac{1}{4}$ ft. We need to use Hooke's Law to find k.

$$9 = \frac{1}{4}k \implies k = 36$$

Therefore, in this problem the function we use is $F(x) = kx$, and $k = 36$, thus

$$F(x) = 36x$$

To determine the amount of work needed to stretch the spring from 6 in to 12 in, we need to consider the fact that the amount of force varies as it is stretched. Hence, we integrate to find the total work. $36x$ is the force, and dx is the distance the spring is stretched. Convert 6 in to $\frac{1}{2}$ ft and 12 in to 1 ft.

$$\text{Work} = \int_{\frac{1}{2}}^{1} F(x)\,dx = \int_{\frac{1}{2}}^{1} 36x\,dx = 18x^2 \Big|_{\frac{1}{2}}^{1} = 18 - \frac{18}{4} = \frac{27}{2} \text{ ft-lb}$$

Summary. The first step in *spring* problems is to apply Hooke's Law to determine k. Then picture stretching the spring little by little. In this problem, the force used to stretch the spring x inches beyond the natural length of 2 inches is $F(x) = 36x$. The distance it is stretched little by little is dx. Then we integrate to add up all the work done.

Be careful that all the units match before integrating.

Example 2: A Leaky Bucket

A 5 lb bucket is lifted up 20 ft from the ground. The bucket initially has 2 gallons (= 16 lb) of water. Water is leaking out at a constant rate. It is half empty when it reaches the top. Find the work done when lifting up the bucket and the water.

Strategy. The *easy* part here is the bucket because its weight remains constant. Because water is leaking out of the bucket, the weight of the water in the bucket varies as it is being lifted. We will separate the work for the constant part, and the variable part.

Solution. The work done to lift the bucket is $W_b = 5\,\text{lb} \cdot 20\,\text{ft} = 100$ foot-pounds.

The force of the water varies depending on the height of the bucket. Initially at height $h = 0$, the force is 16 lb. At height $h = 20$, the force is 8 lb, since half of it has leaked out. The water is leaking out at the rate of $\dfrac{8\,\text{lb}}{20\,\text{ft}}$ or $\dfrac{2}{5}$ pounds per foot.

At height h, $16 - \frac{2}{5}h$ lb of water is left in the bucket, thus the force of the water at height h is

$$F(h) = 16 - \frac{2}{5}h$$

Now, the weight of the water changes while the bucket is lifted up, thus we lift the water up bit by bit a distance of dh. Hence, the work needed to lift the water is:

$$
\begin{aligned}
W_w &= \int_0^{20} \underbrace{F(h)}_{\text{force}}\,\underbrace{dh}_{\text{distance}} \\
&= \int_0^{20} \left(16 - \frac{2}{5}h\right)dh \\
&= \left. 16h - \frac{1}{5}h^2 \right|_0^{20} \\
&= \left(16 \cdot 20 - \frac{1}{5} \cdot 400\right) - (0 - 0) \\
&= 240 \text{ foot-pounds}
\end{aligned}
$$

The total work needed to lift the bucket and the water is $W_b + W_w = 100 + 240 = 340$ foot-pounds.

Summary. In this example the force, which is the combined weight of the bucket and the water, is variable. Therefore, we lifted the bucket little by little. In a *bucket* example, we have

$$\begin{cases} \text{Force} = \text{force of bucket and water} \\ \text{Distance} = dh \end{cases}$$

Example 3: Filling a Tank

A cylindrical tank is sitting 100 ft above the ground and has radius 50 ft and height 20 ft. The density (or weight) of water is $\rho = 62.4\,\text{lb}/\text{ft}^3$.

a) How much work is done in filling this tank, from ground level through the bottom of the tank, with water?

b) How much work is done to fill half of the tank?

c) How much work is done in pumping half of the water out through the top of the tank, assuming the tank is completely filled?

Strategy. We cannot simply find the volume of the tank and lift up all the water at once because the amount of work is different for the water at different levels. For example, it takes more work to lift a *slice* of water all the way to the top of the tank, than it takes to lift it to a lower part of the tank. Since the amount of work varies, we must make horizontal slices, compute the work needed for one slice, let the limit go to an infinite number of slices and then add them all together by integration.

Solution. a) A typical slice of water at height h, with h between 100 and 120 feet, is a disk.

A typical slice of water

The volume of a typical slice is
$$V = \pi r^2 h = \pi \cdot 50^2 \Delta h = 2500\pi \Delta h\,\text{ft}^3$$

Since water weighs 62.4 lb/ft^3, the force (or weight) of a typical *slice* is:
$$F = 62.4\,\text{lb}/\text{ft}^3 \cdot 2500\pi \Delta h\,\text{ft}^3 = 156{,}000\pi \Delta h\,\text{lb}$$

The slice of water is lifted up to level h, thus the distance is h (h is between 100 and 120).

$$\begin{aligned}
\text{Work to fill the tank} &= \int_{100}^{120} \underbrace{h}_{\text{distance}} \cdot \underbrace{156{,}000\pi}_{\text{force}}\,dh \\
&= 78{,}000\pi h^2 \Big|_{100}^{120} \\
&= 78{,}000\pi\,|14{,}400 - 10{,}000| \\
&= 343{,}200{,}000\pi \text{ foot-pounds}
\end{aligned}$$

b) In order to fill half the tank, we need to fill the water to level h, but now h is between 100 and 110.

$$\text{Work to fill half the tank} = \int_{100}^{110} 156{,}000\pi h\, dh$$
$$= 78{,}000\pi h^2 \Big|_{100}^{110}$$
$$= 78{,}000\pi\, (12{,}100 - 10{,}000)$$
$$= 163{,}800{,}000\pi \text{ foot-pounds}$$

c) In order to pump water over the top, each slice is lifted from level h to the top of the tank at 120 ft. The distance in this case is $120 - h$.

$$\text{Work to empty half of the tank} = \int_{110}^{120} \underbrace{(120-h)}_{\text{distance}} \cdot \underbrace{156{,}000\pi\, dh}_{\text{force}}$$
$$= \pi \int_{110}^{120} (18{,}720{,}000 - 156{,}000h)\, dh$$
$$= \pi\, (18{,}720{,}000h - 78{,}000h^2) \Big|_{100}^{120}$$
$$= 7{,}800{,}000\pi \text{ foot-pounds}$$

Summary. Note that the amount of work to fill the tank half full is less than half the work needed to fill the entire tank. This is because the amount of work varies. The less empty the tank is, the higher the water must go, and consequently, it takes more work

There is a fundamental difference between the tank problems, and the previous bucket problem.

For example, in the tank problem, part a), the force of a *slice* of water is constant. Since the force is constant in the tank problem, we can lift a *slice* of water into place.

$$\begin{cases} \text{Force (constant)} = \text{force of a typical slice at height } h = 156{,}000\pi\, dh \\ \text{Distance} = h \end{cases}$$

Compare this to the bucket problem. The force is variable, it changes for each little distance that the bucket is lifted, thus the distance is equal to dh.

$$\begin{cases} \text{Force (variable)} = \text{force of bucket plus water at height } h \\ \text{Distance} = dh \end{cases}$$

Example 4: Mass and Center of Mass

Find the center of mass of the quadrilateral with vertices $(0,0)$, $(0,2)$, $(2,2)$ and $(4,0)$.

Strategy. We will find the area using geometry, and then find the center of mass using the formula

$$\text{Center of mass} = \frac{Moment}{mass}$$

Solution. We first sketch the quadrilateral. We can assume that it is of uniform density. We can think of the center of mass as the balance point of the quadrilateral. See sketch.

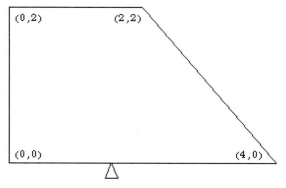

To make computations easier, we assume that the density is one unit per one square unit. The density function at each point x is then given by the height of the quadrilateral at a point x that is

$$\rho(x) = \begin{cases} 2 \text{ for } 0 \leq x \leq 2 \\ 4 - x \text{ for } 2 < x \leq 4 \end{cases}$$

We need to find the Mass and the Moment.

$$\begin{aligned} \text{Mass } m &= \int_0^4 \rho(x)\,dx = \text{net area under } \rho(x) \\ &= \text{area of rectangle plus area of triangle} \\ &= 4 + 2 = 6 \end{aligned}$$

$$\begin{aligned} \text{Moment } M &= \int_0^4 x\rho(x)\,dx \\ &= \int_0^2 2x\,dx + \int_2^4 x(4-x)\,dx \end{aligned}$$

$$= \int_0^2 2x\,dx + \int_2^4 \left(4x - x^2\right) dx$$

$$= x^2 \Big|_0^2 + \left(2x^2 - \frac{1}{3}x^3\right)\Big|_2^4$$

$$= (4 - 0) + \left(32 - \frac{64}{3}\right) - \left(16 - \frac{8}{3}\right)$$

$$= \frac{28}{3}$$

$$\text{Center of Mass} = \frac{M}{m} = \frac{\frac{28}{3}}{6} = \frac{28}{18} = \frac{14}{9} = 1.\overline{5}$$

Summary. We have found the center of mass which tells us that if we placed the quadrilateral at $x = 1.\overline{5}$, then the entire shape would be balanced. See picture.

5.6. Probability

Key Topics.
- Probability Density Function (pdf)
- Mean and Median

Worked Examples.
- Scores on a Test
- Mean and Median of a pdf
- Example of A Normal Distribution

Overview

In this section, we briefly see how calculus is used in the study of probability and statistics. It may surprise you to discover that when you received your percentile ranks on your SAT's, ACT's, GRE's or any other nationally normed test, you were actually given the area under a curve corresponding to your test score. As you know, area under a curve is found by integration.

Probability Density Function (pdf). If X is a random variable which may assume any value x with $a \leq x \leq b$, then a pdf for X is a function $f(x)$ satisfying:

i) $f(x) \geq 0$ for all $a \leq x \leq b$

ii) $\int_a^b f(x)dx = 1$

If we realize that we are measuring probabilities, it makes sense that the function must be positive. Also, since we are considering every possibility, the sum of all the probabilities, the integral, must be 1 or 100%. Nothing else can happen.

If c and d are numbers such that $[c,d]$ is a subset of $[a,b]$, then the probability of X falling between c and d is given by:

$$P(c \leq x \leq d) = \int_c^d f(x)dx$$

Note that the probability corresponds to the area under the curve of the function.

Mean and Median. The mean is actually what we think of when we use the word *average*. If there is a pdf, then the mean denoted by μ (mu) can be found by:

$$\mu = \int_a^b xf(x)dx$$

Another type of average used by statisticians is the median. This is the number which exactly divides the probability in half. Think about the median line which runs right down the middle of the highway. Half of the values lie above the median and half of the values lie below the median. Depending on the scenario, the mean or the median together gives a better description of what is actually happening. In order to find the median, solve the following integral for the constant c.

$$\int_0^c f(x)dx = .5$$

Example 1: Scores on a Test

The graph below shows the probability of scores on a calculus test.

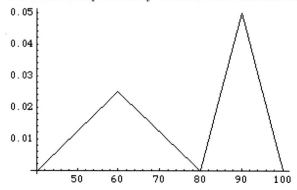

a) Check to see that this function is a probability density function (pdf).

b) What is the probability that a student receives a score between 40 and 80 points?

c) What is the probability that a student receives a score between 90 and 100 points?

d) Find the median and the mean of this pdf.

Strategy. We are able to obtain most of what we need just by looking at the graph. To obtain the mean, we will have to use technology.

Solution. a) In order to determine that this function is a pdf, we need to be sure that $f(x) \geq 0$, and that $\int_0^{100} f(x)dx = 1$. Note that the values of f are all positive, and

$$\begin{aligned}\int_0^{100} f(x)dx &= \text{area of the two triangles} \\ &= \frac{1}{2} \cdot 40 \cdot 0.025 + \frac{1}{2} \cdot 20 \cdot 0.05 \\ &= 0.5 + 0.5 = 1\end{aligned}$$

b) The probability that a student receives a score between 40 and 80 points is given by

$$P(40 < X < 80) = \int_{40}^{80} f(x)dx = \frac{1}{2} \cdot 40 \cdot 0.025 = 0.5 = 50\%$$

c) The probability that a student receives a score between 90 and 100 points is given by

$$P(90 < X < 100) = \int_{90}^{100} f(x)dx = \frac{1}{2} \cdot 10 \cdot 0.05 = 0.25 = 25\%$$

d) The median in this example is easy to find by looking at the graph. Half of the scores are above 80, and half are below 80. The area of both triangles are 0.5 = 50%. Therefore the median is 80.

The mean μ is more difficult to find. We have to find the defining equation of the function $y = f(x)$.

$$f(x) = \begin{cases} \frac{0.025}{20}(x-40) & \text{for } 40 \leq x \leq 60 \\ \frac{-0.025}{20}(x-80) & \text{for } 60 \leq x \leq 80 \\ \frac{0.05}{10}(x-80) & \text{for } 80 \leq x \leq 90 \\ \frac{-0.05}{10}(x-40) & \text{for } 90 \leq x \leq 100 \end{cases}$$

We use a CAS to find the integrals in this example.

$$\begin{aligned}
\mu &= \int_0^{100} x f(x) dx \\
&= \int_{40}^{60} x \cdot \frac{0.025}{20}(x-40)\, dx + \int_{60}^{80} x \cdot \frac{-0.025}{20}(x-80)\, dx \\
&\quad + \int_{80}^{90} x \cdot \frac{0.05}{10}(x-80)\, dx + \int_{90}^{100} x \cdot \frac{-0.05}{10}(x-40)\, dx \\
&= \frac{40}{3} + \frac{50}{3} + \frac{65}{3} + \frac{70}{3} = \frac{225}{3} = 75
\end{aligned}$$

Thus the mean is 75, which is slightly below the median of 80.

Summary. In this example, we were able to answer all but one question by geometrically finding areas under the probability density function. The mean is more difficult to find in cases where the graph of the pdf is given, but no formula.

Example 2: Mean and Median of a pdf

Find the mean and median of the random variable with pdf $f(x) = \frac{1}{x}, 1 \leq x \leq e$.

Strategy. First we will make sure that we actually have a pdf, then we will be able to use the formulas to obtain the mean and median.

Solution. We first check to make sure that f is a pdf. $f(x) = \dfrac{1}{x}$, is always positive on the given domain. Also, $\int_1^e \dfrac{dx}{x} = \ln x \, \big|_1^e = \ln e - \ln 1 = 1$.

$$\begin{aligned}\text{Mean } \mu &= \int_1^e x f(x)\,dx = \int_1^e x \cdot \frac{1}{x}\,dx = \int_1^e dx \\ &= x \, \big|_1^e = e - 1\end{aligned}$$

To find the median c, we need to solve

$$\int_1^c f(x)\,dx = 0.5$$

$$\int_1^c \frac{dx}{x} = \ln x \, \big|_1^c = \ln c - \ln 1 = \ln c$$

$$\implies \ln c = 0.5 \implies c = e^{0.5} = \sqrt{e}$$

Thus the mean is $e - 1 \approx 1.72$, and the median is $\sqrt{e} \approx 1.65$.

Summary. Once you are certain that you have a pdf, then you are free to use the relatively simple formulas for the mean and the median. This has given you a small taste of statistics. You will learn a lot more when you take an entire course in statistics.

Example 3: Example of A Normal Distribution

Suppose the annual rainfall in inches in Calc Town is distributed normally with normal distribution

$$f(x) = \frac{1}{3\sqrt{2\pi}} e^{-\frac{(x-20)^2}{2 \cdot 3^2}}$$

What is the probability that the annual rainfall will be between 17 and 23 in ?

Strategy. A normal distribution is by definition a pdf. All we need to do in order to find the probability that the amount of rainfall is between 17 and 23 inches is to integrate the function between those limits.

Solution. The probability of the annual rainfall being between 17 and 23 inches is given by

$$P(17 < X < 23) = \int_{17}^{23} \frac{1}{3\sqrt{2\pi}} e^{-\frac{(x-20)^2}{2 \cdot 3^2}} dx$$

This is an integral that we cannot integrate with our techniques. We use a CAS to evaluate this integral and find

$$P(17 < X < 23) \approx 0.68 = 68\%$$

This example illustrates an interesting fact. This normal distribution has mean $\mu = 20$, and standard deviation $\sigma = 3$. Since we were looking at numbers from 17 to 23, we were looking at the amount of rainfall being one standard deviation from the mean: 20 ± 3. The probability of something being within one standard deviation from the mean on a normally distributed function is always .68 or 68%.

Summary. Once again, as long as we are certain that we have a pdf, we can use the formulas associated with pdfs. In this problem, we used the fact that to find the probability of a subset of the entire interval, we only need to integrate over the smaller interval.

CHAPTER 6

Differential Equations

6.1. Growth and Decay Problems

Key Topics.
- Exponential Growth and Decay
- Compound Interest

Worked Examples.
- A Fox Population
- Half-Life of Radium
- Newton's Law of Cooling for an Apple Pie
- Compound Interest

Overview

In this section we use calculus to develop the formulas for exponential growth and decay that were used in college algebra and precalculus. We will work several examples using these formulas. However, in this section, the problems only require precalculus knowledge. It's the development of the formulas which utilize methods of calculus.

Exponential Growth and Decay. Exponential growth and decay is used for problems concerning population growth and radioactive decay. The set-up of the problem comes from the statement that the rate of change is proportional to the amount present. This translates symbolically to $y' = \dfrac{dy}{dx} = ky$ where k is the constant of proportionality. Note that we use $y' = \dfrac{dy}{dx}$ as the translation of *rate of change*.

We can solve this differential equation by a process called separation of variables, which is covered more thoroughly in the next section. Once we solve this equation, then we can use the resulting formula for all problems of this type without having to solve the differential equation over again each time.

$$\begin{aligned} \dfrac{dy}{dt} &= ky \\ \dfrac{1}{y}dy &= k\,dt \end{aligned}$$

6.1. GROWTH AND DECAY PROBLEMS

$$\int \frac{1}{y} dy = \int k \, dt$$
$$\ln y = kt + c_1$$

By the definition of logarithms, we can solve for y as

$$y = e^{kt+c_1}$$

By properties of exponents, we add the exponents when we are multiplying with the same base. Thus, we can rewrite the equation as

$$y = e^{kt} \cdot e^{c_1}$$

Note that e^{c_1} is a constant. We can call this constant A, and then we have the formula

$$y = Ae^{kt}$$

In examples of growth, k will be positive, in examples of decay, k will be negative.

Compound Interest. Interest grows proportionally to the amount present, also. Consider the fact that the more money you invest, the greater the change in amount when the interest is added. Therefore, compound interest satisfies the same formula $y = Ae^{kt}$.

Often, this is expressed using letters, which more accurately portray what is happening in the problem. We can use $P = P_0 e^{rt}$ where P is the ending amount of money, P_0 is the principal, or the amount of money which was originally invested or borrowed, r is the rate of interest and t is the time in **years**. Many students find this formula easy to remember by calling it the *Pert* formula. Notice that remembering *Pert* works for all exponential growth and decay functions also.

Example 1: A Fox Population

Suppose a population of foxes initially has 20 foxes and doubles every 5 years.
a) Determine the continuous growth rate, and
b) determine when the population will reach 60 foxes.

Strategy. Since we know that the fox population is growing exponentially, and we just found the formula for exponential growth, we can simply use the formula.

Solution. We know that the fox population is growing exponentially. If $y(t) =$ fox population t years from now then $y = Ae^{kt}$. The initial population is 20, and therefore, we know that when $t = 0$, $y = 20$.

$$20 = Ae^0 \implies A = 20$$

Our equation is then

$$y = 20e^{kt}$$

Next, we need to find the constant, k. We know that the population doubles after 5 years. Therefore, $y(5) = 40$. Substituting $t = 5$ and $y = 40$ gives

$$\begin{aligned} 40 &= 20e^{5k} \\ 2 &= e^{5k} \\ \ln 2 &= 5k \\ k &= \frac{1}{5}\ln 2 \approx .1386 \end{aligned}$$

a) Now our equation is $y(t) = 20e^{(.1386)t}$ telling us that the continuous growth rate is approximately 0.14 or 14% per year.

b) To determine when the population will reach 60 foxes, we let $y = 60$ and solve for t.

$$\begin{aligned} 60 &= 20e^{.1386t} & \\ 3 &= e^{.1386t} & \text{(apply ln)} \\ \ln 3 &= \ln e^{.1386t} & \text{(exponents } \textit{jump over ln}\text{)} \\ \ln 3 &= .1386t \ln e & \\ \ln 3 &= (.1386t)(1) & \\ t &= \frac{\ln 3}{.1386} & \\ t &\approx 7.9 & \end{aligned}$$

The population of foxes reaches 60 foxes in approximately 7.9 years.

Summary. Now that we have derived the equation to use for exponential growth and decay, solving these problems involves replacing the variables with the correct values and solving for the unknown variable. It should be a good review of college algebra.

Example 2: Half-Life of Radium

The half-life of radium-226 is 1590 years. If initially there are 200 mg, when does the amount drop below 80 mg?

Strategy. This is an example of exponential decay. The formula is the same as the one in example 1 which we derived. The difference is that the constant, k, will be negative.

Solution. Since the initial amount is 200 mg, we know that $A = 200$ in the equation $y = Ae^{kt}$. We can write $y = 200e^{kt}$.

In addition, we know the half-life, which tells us how long it takes until half of the radium remains. The half-life is given as 1590 years. If half of the initial amount remains, then 100 mg remains after 1590 years. Our equation is now

$$100 = 200e^{k(1590)}$$

We can solve for k.

$$\frac{1}{2} = e^{1590k}$$
$$\ln \frac{1}{2} = 1590k$$
$$k = \frac{\ln \frac{1}{2}}{1590} \approx -.000436$$
$$\implies y(t) = 200e^{-.000436}$$

When is 80 mg left? Let $y = 80$ and solve the resulting equation.

$$80 = 200e^{-.000436t}$$
$$0.4 = e^{-.000436t}$$
$$\ln 0.4 = -.000436t$$
$$t = \frac{\ln 0.4}{-.000436} \approx 2102 \text{ years}$$

Summary. Without using any rounding, and using the fact that "exponents jump over logs", we can rewrite $y(t)$ as follows:

$$y(t) = 200e^{\frac{\ln \frac{1}{2}}{1590}t} = 200e^{\frac{t}{1590}\ln \frac{1}{2}} = 200e^{\ln\left(\left(\frac{1}{2}\right)^{\frac{t}{1590}}\right)}$$

$$= \underbrace{200}_{\text{initial quantity}} \cdot \underbrace{\left(\frac{1}{2}\right)}_{\text{half-life}} \overbrace{^{\frac{t}{1590}}}^{\text{every 1590 years}}$$

Is this the same as the above answer of $y(t) = 200e^{-.000436}$? Yes, because

$$e^{-.000436} = e^{\frac{\ln \frac{1}{2}}{1590}} = e^{\left(\left(\ln \frac{1}{2}\right) \cdot \frac{1}{1590}\right)} = \left(e^{\left(\ln \frac{1}{2}\right)}\right)^{\frac{1}{1590}} = \left(\frac{1}{2}\right)^{\frac{1}{1590}}$$

Example 3: Newton's Law of Cooling for an Apple Pie

An apple pie is placed inside a 375 °F oven. The initial temperature of the pie is 75 °F, and after 10 minutes the temperature has reached 225 °F. Find the temperature at any time t, and then determine how long it will take for the pie to reach a temperature of 350 °F.

Strategy. We use Newton's Law of Heating and Cooling. The formula is derived in the text and here we just state the conclusion.

$$y(t) = Ae^{kt} + T_a$$

where A and k are constants, T_a is the temperature of the surrounding area (called ambient temperature), and t is time.

Solution. We know that the initial temperature is 75 °F and that the temperature after 10 minutes is 225 °F.

This tells us that $T(0) = 75\,°F$, and $T(10) = 225\,°F$ from which we can find the constants A and k.

$T_a = 375\,°F$ because are putting the pie in the oven. First we find A:

$$75 = Ae^0 + 375 \implies A = 75 - 375 = -300$$

$T(10) = 225$ yields

$$\begin{aligned} 225 &= -300e^{10k} + 375 \\ -150 &= -300e^{10k} \\ \frac{1}{2} &= e^{10k} \\ \ln \frac{1}{2} &= 10k \\ k &= \frac{1}{10}\ln\frac{1}{2} \approx -.0693 \end{aligned}$$

a) Thus the temperature of the pie is given by

$$y(t) = -300e^{-.0693t} + 375$$

b) The pie reaches a temperature of 350 °F when

$$\begin{aligned} 350 &= -300e^{-.0693t} + 375 \\ -25 &= -300e^{-.0693t} \\ \frac{1}{12} &= e^{-.0693t} \\ \ln \frac{1}{12} &= -.0693t \\ t &= \frac{\ln \frac{1}{12}}{-.0693} \approx 35.86 \end{aligned}$$

After approximately 36 minutes the pie reaches a temperature of 350 °F.

Summary. Once again we substituted the values for the variables in our equation. We need to first solve for the constants using the given information. Only then, can we do the computation needed to answer the question. Often, there are several steps necessary when there is more than one variable to find.

Example 4: Compound Interest

Suppose that two sisters are each investing $2,000. One sister receives 6% interest compounded annually, and the other 5.9% compounded continuously. How much money will each sister have at the end of a ten year period? Which sister made the better decision?

Strategy. The formula for interest compounded n times a year with an annual interest rate of r, is

$$B = P(1 + \frac{r}{n})^{nt}$$

For interest being compounded continuously, we have the Pe^{rt} formula:

$$B = Pe^{rt}$$

Solution. Let's call the balance for sister #1, B_1. She invested her $2000, so that the interest was compounded annually, that is one time per year. The amount of money she has at the end of ten years is:

$$\begin{aligned} B_1 &= 2000\left(1 + \frac{.06}{1}\right)^{1(10)} \\ &= 2000(1.06)^{10} \\ &\approx \$3581.70 \end{aligned}$$

Similarly, let's call the balance for sister #2, B_2. She invested her $2000, so that the interest was compounded continuously. The amount of money that she has at the end of a ten year period is:

$$\begin{aligned} B_2 &= 2000e^{.059(10)} \\ &\approx \$3607.98 \end{aligned}$$

Obviously, the second sister who invested her money to be compounded continuously at 5.9% interest made the better decision.

Summary. As we see here, even though the rate of interest was lower, the fact that it was continuously compounded made the final amount almost $100 greater. It shows that we need to really think about and understand interest rates when we are investing or borrowing money.

6.2. Separable Differential Equations

Key Topics.
- Separable Equations
- Logistic Equation

Worked Examples.
- Solving a Separable Equation
- Solving an Initial Value Problem
- Solving a Logistic Growth Problem

Overview

In this section, we begin the study of differential equations. If an equation contains a derivative, then it is called a differential equation and we are looking for a **function** to satisfy that equation. If the equation contains only the first derivative, then we call it a first order differential equation.

Separable Equations. A differential equation is separable if we can separate the variables through multiplication or division so that one of the variables is on one side and the other variable is on the other side. Then we will integrate both sides to find the solution.

Logistic Equation. If there is a maximum sustainable population, then there is a constraint on the growth of the population. Also, as the population approaches its maximum, the rate of growth slows. The rate of change then is not only proportional to the present value as it was in section 6.1, but it is jointly proportional to the present population **and** the difference between the current level and the maximum, M. Then M is called the carrying capacity.

$$y'(t) = ky(M-y)$$

The solution to this differential equation is

$$y = \frac{AMe^{kMt}}{1 - Ae^{kMt}}$$

where A is a constant.

Example 1: Solving a Separable Equation

Find the general solution of the separable equation $\frac{dy}{dx} = -\frac{4x}{9y}$, and sketch several members of the family of solutions.

Strategy. We need to separate the variables so that the $9y$ is on the same side with dy and the $-4x$ is on the same side with dx. Also, the dy and the dx must stay in the numerator. We can never integrate an expression similar to $\frac{1}{dx}$.

Once the variables are separated, we integrate both sides. We do not need to put $+c$ on both sides as we can put $+c$ on the right side only which will represent a combination of the two constants of integration.

Solution.

$$\frac{dy}{dx} = \frac{-4x}{9y} \qquad \text{(separate variables)}$$

$$9y\,dy = -4x\,dx$$

$$\int 9y\,dy = \int -4x\,dx$$

$$\frac{9}{2}y^2 = -2x^2 + c \qquad \text{(only one } c \text{ is needed)}$$

$$2x^2 + \frac{9}{2}y^2 = c$$

If $c > 0$ then this gives an ellipse. Therefore the family of solutions are ellipses.

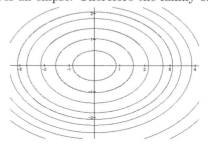

Summary. The differential equation is considered solved because we have a relationship between x and y. If desired, we can continue and solve for y as:

$$\frac{9y^2}{2} = -2x^2 + c_1$$

$$9y^2 = -4x^2 + c_2 \qquad (c_2 = 2c_1)$$

$$y^2 = \frac{-4}{9}x^2 + c \qquad (c = \tfrac{1}{9}c_2)$$

$$y = \pm\sqrt{-\frac{4}{9}x^2 + c}$$

Example 2: Solving an Initial Value Problem

Find the solution of the initial value problem

$$\frac{dy}{dx} = x^2 y \text{ with } y(0) = 10$$

Strategy. We solve this differential equation using separation of variables similar to the previous example. In this problem we have one additional step, we can actually find a value for c by using the information that $y(0) = 10$.

Solution.

$$\begin{aligned}
\frac{dy}{dx} &= x^2 y \\
\frac{1}{y} dy &= x^2 dx \\
\int \frac{1}{y} dy &= \int x^2 dx \\
\ln|y| &= \frac{1}{3}x^3 + c_1 && \text{(by definition of ln)} \\
|y| &= e^{\frac{1}{3}x^3 + c_1} && (e^{a+b} = e^a e^b) \\
|y| &= e^{\frac{1}{3}x^3} \cdot e^{c_1} && (C = e^{c_1} = \text{constant}) \\
|y| &= Ce^{\frac{1}{3}x^3}
\end{aligned}$$

$y(0) = 10$ implies that $10 = Ce^0 \implies C = 10$.
Since the initial value of y is positive, $|y| = y$.
Thus the solution to the differential equation is

$$y = 10 e^{\frac{1}{3}x^3}$$

Summary. This is the same method we used to derive the formula for continuous growth and decay at the beginning of the last section (6.1). Make sure that you understand why $e^{f(x)+c_1}$ can be rewritten as $Cf(x)$.

Example 3: Solving a Logistic Growth Problem

Suppose a fox population follows the logistic model given by the differential equation:

$$\frac{dP}{dt} = P(1000 - P)$$

Solve the differential equation by separating variables, and check your answer by using the formula for logistic growth from the text. Also, determine the long term effect if the initial population is 500.

(In section 6.3 we will continue this problem by looking at slope fields)

Strategy. We are going to actually solve this equation as opposed to simply using the formula from the book. Then we will show that our answer and the book's formula are equivalent.

Solution. $\frac{dP}{dt} = P(1000 - P)$ is a separable equation. Separating variables gives

$$\int \frac{dP}{P(1000 - P)} = \int dt$$

We use partial fractions to solve the integral on the left hand side.

$$\frac{1}{P(1000 - P)} = \frac{A}{P} + \frac{B}{1000 - P}$$
$$1 = A(1000 - P) + BP$$

This equation holds for all values of P, in particular for $P = 0$, and $P = 1000$.

$$P = 0 \implies 1 = 1000A \implies A = \frac{1}{1000}$$

$$P = 1000 \implies 1 = 1000B \implies B = \frac{1}{1000}$$

Thus

$$\frac{1}{P(1000 - P)} = \frac{\frac{1}{1000}}{P} + \frac{\frac{1}{1000}}{1000 - P} = \frac{1}{1000}\left(\frac{1}{P} + \frac{1}{1000 - P}\right)$$

Before we integrate the above separable equation, note that one of the integrals will be $\int \frac{dP}{1000 - P}$. Use the substitution $u = 1000 - P$, then $-du = dP$, and we have

$$\int \frac{dP}{1000 - P} = -\int \frac{du}{u} = -\ln|u| + c = -\ln|1000 - P| + c$$

Now, we continue with the separable equation from above. Assume that the number of foxes is between 0 and 1000 so that $1000 - P > 0$.

$$\int \frac{dP}{P(1000-P)} = \int dt$$

$$\frac{1}{1000}\int \left(\frac{1}{P} + \frac{1}{1000-P}\right) dP = \int dt$$

$$\frac{1}{1000}\left(\ln|P| - \ln|1000-P|\right) = t + c_1$$

$$\ln\frac{P}{1000-P} = 1000t + c_2 \qquad (c_2 = 1000c_1)$$

$$e^{1000t+c_2} = \frac{P}{1000-P}$$

$$e^{1000t} \cdot e^{c_2} = \frac{P}{1000-P} \quad (A = e^{c_2} = \text{constant})$$

$$\frac{P}{1000-P} = Ae^{1000t}$$

Now comes the difficult part of solving the last equation for P.

$$\frac{P}{1000-P} = Ae^{1000t}$$
$$P = (1000-P)Ae^{1000t}$$
$$P = 1000Ae^{1000t} - PAe^{1000t}$$
$$P + PAe^{1000t} = 1000Ae^{1000t}$$
$$P\left(1 + e^{1000t}\right) = 1000Ae^{1000t}$$
$$P = \frac{1000Ae^{1000t}}{1 + Ae^{1000t}}$$

We can check to make sure that we have the correct answer by using the formula in the text. In this example $M = 1000$ is the carrying capacity, and $k = 1$ is the growth rate.

$$y = \frac{AMe^{kMt}}{1 + Ae^{kMt}} = \frac{1000Ae^{1000t}}{1 + Ae^{1000t}} \;\checkmark$$

Finally, if the initial population is 500, that is when $t = 0$, $P = 500$. From this we can find A.

$$500 = \frac{1000Ae^0}{1 + Ae^0} = \frac{1000}{1+A}$$
$$\implies 500 + 500A = 1000A$$
$$\implies 500 = 500A$$
$$A = 1$$

The population of foxes is given by

$$P = \frac{1000e^{1000t}}{1 + e^{1000t}} \text{ foxes}$$

The carrying capacity is $M = 1000$, that implies that the population over time will approach 1000 foxes. We can see this nicely from a slope field in the next section.

Summary. The carrying capacity is $M = 1000$, that implies that the population over time will approach 1000 foxes. We will be able to see this nicely when we learn about slope fields in the next section.

6.3. Direction Fields and Euler's Method

Key Topics.
- Direction Fields
- Equilibrium Solutions
- Euler's Method

Worked Examples.
- The Direction Field of a Logistic Growth Example
- Euler's Method

Overview

Unfortunately, most differential equations are not solved as easily as the ones we just solved using separation of variables. In this section, we use direction fields and Euler's method to approximate solutions.

Direction Fields. Since we are given the derivative of the function we are looking for, called the solution function, we know the **slope** of that function. We can sketch the slope of this function at lots of points making very small arrows in the direction of the slope. In this way we have a picture and hence, a general idea of what the function looks like.

Equilibrium Solutions. An equilibrium solution is a constant solution. The rate of change that is the derivative, therefore, is zero, telling us that there is no change at that point. The example, in the text talks about an ant infestation. The equilibrium solution is where the ant population is neither increasing nor decreasing.

Euler's Method. Euler's method enables us to approximate a single solution curve. This is a lot less *messy* than looking at a direction field. It is reminiscent of using Newton's method for approximating the zeros of a function. We use the tangent line approximations just as we did with Newton's method. We use a *step* of h units to go from one x-value to another.

Starting with the differential equation
$$y' = f(x, y)$$
at a point (x_0, y_0), we use the tangent line at that point (which we can easily obtain since we know the derivative and hence, the slope) to approximate the next point (x_1, y_1) along the solution curve. If $x_1 = x_0 + h$, then
$$y(x_1) \approx y_1 = y_0 + y'(x_0)(x_1 - x_0)$$

Since $x_1 - x_0 = h$, and $y'(x_0) = f(x_0, y_0)$, we have $y_1 = y_0 + h \cdot f(x_0, y_0)$.

Continue in this manner, repeating all the steps using (x_1, y_1) in place of (x_0, y_0) and obtaining (x_2, y_2) etc. In general, then, we can say:
$$y(x_{i+1}) \approx y_{i+1} = y_i + h \cdot f(x_i, y_i)$$

Example 1: The Direction Field of a Logistic Growth Example

Suppose a fox population P (measured in thousands of foxes) follows the logistic model given by the differential equation $\dfrac{dP}{dt} = P(1000 - P)$. The slope field is given below. Sketch solutions for $P_0 = 500$, $P_0 = 1000$, $P_0 = 1200$, and $P_0 = -200$ (ignore that a fox population cannot be negative). Find the equilibrium solutions and determine the stability of all equilibrium solutions. What are the long term trends?

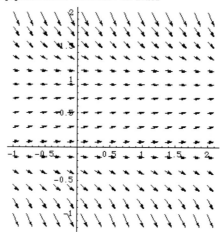

Strategy. First we will follow the flow of the tangent lines in the slope field and sketch the solutions. Note that the scale of the vector field is given in thousands of foxes. We will then find the equilibrium solutions. From there, we will decide from the direction field what is happening around the equilibrium solutions, and finally draw our conclusions about long term trends.

Solution.

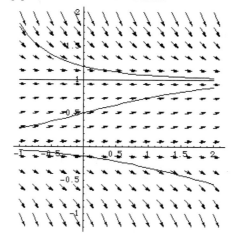

Next we find the equilibrium solutions. These are the constant solutions where $\frac{dP}{dt} = 0$. Here

$$\frac{dP}{dt} = P(1000 - P) = 0 \implies P = 0 \text{ or } P = 1000$$

Summary. From the slope field we can see that $P = 0$ is an unstable solution since the solutions close to $P = 0$ are going in a direction further away from 0. The solution $P = 1000$ is stable, since all solutions close to this equilibrium solution tend to approach this solution. Therefore, the trend in this population is to approach a population size of 1000, also called the **limiting** or **carrying value** of this population.

Example 2: Euler's Method

Find an approximate solution for the initial value problem $y' = x^2 - y^2$ on the interval $[0, 1]$ with $y(0) = 1$.

Strategy. Let's choose $h = 0.2$ and start at $x_0 = 0$. It will be easy to identify the x_i's because we just continually add $h = 0.2$. We will use Euler's method to obtain the y-values.

Solution. The x_i's are easy to find.
$x_1 = 0.2$, $x_2 = 0.4$, $x_3 = 0.6$, $x_4 = 0.8$, and $x_5 = 1$.

The y-values take a little more effort. Euler's method uses the formula $y(x_{i+1}) \approx y_{i+1} = y_i + hf(x_i, y_i)$ to find the y_i's. We have $f(x,y) = x^2 - y^2$, and starting with $y_0 = 1$, we get

$$\begin{aligned} y_1 &= y_0 + hf(x_0, y_0) \\ &= 1 + 0.2\left(0^2 - 1^2\right) \\ &= 0.8 \end{aligned}$$

$$\begin{aligned} y_2 &= y_1 + hf(x_1, y_1) \\ &= 0.8 + 0.2\left(0.2^2 - 0.8^2\right) \\ &= 0.68 \end{aligned}$$

$$\begin{aligned} y_3 &= y_2 + hf(x_2, y_2) \\ &= 0.68 + 0.2\left(0.4^2 - 0.68^2\right) \\ &= 0.61952 \end{aligned}$$

$$\begin{aligned} y_4 &= y_3 + hf(x_3, y_3) \\ &= 0.61952 + 0.2\left(0.6^2 - 0.61952^2\right) \\ &= 0.6147589939 \end{aligned}$$

$$\begin{aligned} y_5 &= y_4 + hf(x_4, y_4) \\ &= 0.6147589939 + 0.2\left(0.8^2 - 0.6147589939^2\right) \\ &= 0.6671732698 \end{aligned}$$

The solution is sketched below in the slope field.

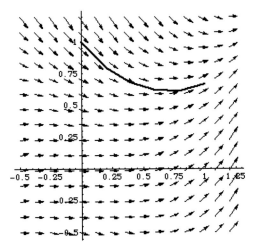

Summary. Although Euler's method can get tedious, it does give us a way to approximate the solution when we do not have another way to obtain it.

6.4. Second Order Equations with Constant Coefficients

Key Topics.
- Second Order Equations with Constant Coefficients

Worked Examples.
- Finding General Solutions
- Initial Value Problem

Overview

In this section we learn how to solve second order differential equations with constant coefficients. A second order differential equation contains a second derivative and may or may not contain the first derivative and the function itself. You can compare this to a quadratic equation which must contain a variable to the second power, and may or may not contain the variable to the first power and a constant.

Second Order Equations with Constant Coefficients. To find the solution of a second order differential equation $ay''(t) + by'(t) + cy(t) = 0$, there are three models to follow. First find the roots of the characteristic equation $ar^2 + br + c = 0$. They will fall into one of the three cases which follow. Use these cases as the models once you decide which type of roots there are.

Case 1: The roots r_1 and r_2 are two distinct real roots. In this case the general solution to the differential equation is $y(t) = c_1 e^{r_1 t} + c_2 e^{r_2 t}$.

Case 2: The characteristic equation has only one root r, i.e. a repeated root. In this case the general solution is $y(t) = c_1 e^{rt} + c_2 t e^{rt}$.

Case 3: The characteristic equation has complex roots $r_1 = u + vi$, and $r_2 = u - vi$. In this case the general solution is $y(t) = c_1 e^{ut} \cos(vt) + c_2 e^{ut} \sin(vt)$.

Example 1: Finding General Solutions

Find the general solutions of the equations
a) $y'' - 5y' + 6y = 0$
b) $y'' - 2y' + y = 0$
c) $y'' - 2y' + 2y = 0$

Strategy. In each case, we will first solve the characteristic equation $ar^2 + br + c = 0$. Then, depending of the type of roots, we will decide which model to use.

Solution. a) The characteristic equation $r^2 - 5r + 6 = 0$ can easily be solved by factoring this equation into $(r-2)(r-3) = 0$, which leads to the solutions $r = 2$, and $r = 3$. This is an example of case 1 and the general solution is:

$$y(t) = c_1 e^{2t} + c_2 e^{3t}$$

b) The characteristic equation $r^2 - 2r + 1 = 0$ can also be solved by factoring. $(r-1)(r-1) = 0$ which leads to the solution of one repeated zero $r = 1$. This is an example of case 2, and the general solution is

$$y(t) = c_1 e^t + c_2 t e^t$$

c) The characteristic equation $r^2 - 2r + 2 = 0$ must be solved using the quadratic formula. There are two complex zeros $r = 1 - i$, and $r = 1 + i$. This is an example of case 3 and the general solution is:

$$y(t) = c_1 e^t \cos t + c_2 e^t \sin t$$

Summary. Note that we first write the characteristic equation of the differential equation, which gives us a quadratic equation. Once, that is done, we solve the quadratic equation and decide which model to use to write the function $y(t)$.

Example 2: Initial Value Problem

Find the solution of the initial value problem $y'' - 8y' + 15y = 0$, $y(0) = 0$, and $y(1) = e^3$.

Strategy. We solve this problem exactly the same way we solved the problems in the previous example. After we write the function $y(t)$, there will be two constants. We can use the two initial values to solve for the two constants.

Solution. Solving the characteristic equation $r^2 - 8r + 15 = 0$ by factoring we find two distinct real zeros at $r = 3$ and $r = 5$. This is another example of case 1, and, therefore, the general solution is
$$y(t) = c_1 e^{3t} + c_2 e^{5t}$$
The initial conditions $y(0) = 0$ gives
$$\text{i) } 0 = c_1 + c_2 \implies c_2 = -c_1$$
And $y(1) = e^3$ gives
$$\text{ii) } e^3 = c_1 e^3 + c_2 e^5$$
Substituting $c_2 = -c_1$ from equation i) into equation ii) leads to
$$e^3 = c_1 e^3 - c_1 e^5 = c_1 \left(e^3 - e^5\right) \implies \begin{cases} c_1 = \frac{e^3}{e^3 - e^5} = \frac{1}{1 - e^2} \\ c_2 = \frac{-1}{1 - e^2} \end{cases}$$
The solution to this initial value problem is:
$$y = \frac{1}{1 - e^2} e^{3t} - \frac{1}{1 - e^2} e^{5t}$$
$$y = \frac{1}{1 - e^2} \left(e^{3t} - e^{5t}\right)$$

Summary. Solving these differential equations with given initial values, makes us solve quadratic equations and systems of equations. Also, we get to do a lot of work using the exponential function. In other words, we are getting an excellent review of algebraic concepts.

6.5. Nonhomogeneous Equations: Undetermined Coefficients

Key Topics.
- Nonhomogeneous Equations

Worked Examples.
- An Example of a Nonhomogeneous Equation

Overview

In this section we expand our solving of second order differential equations. In the previous section we found solutions to equations of the form $ay''(t) + by'(t) + cy(t) = 0$. In this section the right hand side is not equal to 0, instead it is equal to a function $f(t)$.

Nonhomogeneous Equations. When the right hand side of the second order differential equation is 0, it is called a **homogeneous** equation. If the right hand side is a function of t, $F(t) \neq 0$, then it is called a **nonhomogeneous** equation. The equation is of the form $mu'' + cu' + ku = F(t)$. We begin solving it by writing 0 in place of $F(t)$ and using the same techniques as in the previous section. The solution we obtain for $mu'' + cu' + ku = 0$ can be written as $u = c_1 u_1 + c_2 u_2$.

We then need to make an initial guess as to what might be a solution to the equation
$$mu'' + cu' + ku = F(t)$$

There is an excellent table in the text on page 568, which tells us how to choose this initial guess. If we call this guess u_p, then the general solution to the nonhomogeneous differential equation is
$$y = u + u_p$$

Example 1: An Example of a Nonhomogeneous Equation

Find the general solution of the equation

$$y'' - 3y' + 2y = 6e^{4t}$$

Strategy. We first find the roots of the characteristic equation $r^2 - 3r + 2r = 0$. Then we use techniques from section 6.4 to find the general solution u to the equation $u'' - 3u' + 2u = 0$. Finally we will make an educated guess for a solution, u_p, of $y'' - 3y' + 2y = 6e^{4t}$. The general solution is then $y = u + u_p$.

Solution. By factoring, we can find that the roots of the characteristic equation $r^2 - 3r + 2r = 0$ are $r = 1$, and $r = 2$. Therefore, we have two unique real roots which corresponds to the first case in the previous example. The general solution to $u'' - 3u' + 2u = 0$ is, therefore, $u = c_1 e^t + c_2 e^{2t}$.

Now we just need to find one solution of the equation $y'' - 3y' + 2y = 6e^{4t}$. A good guess would be $u_p = Ae^{4t}$, since $F(t) = 6e^{4t}$. Let's see what happens when we try this guess in our differential equation.

$$\begin{aligned} u_p &= Ae^{4t} \\ u_p' &= 4Ae^{4t} \\ u_p'' &= 16Ae^{4t} \end{aligned}$$

$$\begin{aligned} u'' - 3u' + 2u &= 6e^{4t} \\ 16Ae^{4t} - 3 \cdot 4Ae^{4t} + 2Ae^{4t} &= 6e^{4t} \\ 6Ae^{4t} &= 6e^{4t} \implies A = 1 \implies u_p = e^{4t} \end{aligned}$$

Therefore the general solution is

$$\begin{aligned} y &= u + u_p \\ y &= c_1 e^t + c_2 e^{2t} + e^{4t} \end{aligned}$$

Summary. Study the table on page 568 in the text, and try to make sense out of how to tell by looking at $F(t)$ what to choose for the initial guess. In that way, it will be easier to remember and it will also make the whole process more understandable.

CHAPTER 7

Infinite Series

7.1. Sequences of Real Numbers

Key Topics.
- Sequences
- Limit of Sequences
- Bounded Sequences

Worked Examples.
- Finding Limits of Sequences
- Using L'Hôpital's Rule to Find a Limit
- Applying the Squeeze Theorem to A Sequence
- A Monotonic and Bounded Sequence

Overview

In chapter 7 we study sequences and series. More than half of this chapter is devoted to convergence and divergence of series. We will continually be asking whether a series converges to a finite number or not. If not, we say it diverges.

Sequences. The word sequence is used in mathematics just as it is in the English language. A sequence of events is reported in the order in which the events occurred. Similarly, in mathematics, the term sequence refers to an infinite collection of real numbers, written in a specific order. More explicitly, a sequence is a function whose domain is the set of non-negative integers.

Limit of Sequences. If $\lim_{n \to \infty} a_n = L$, where L is a finite number, then we say that the limit of the sequence $\{a_n\}_{n=1}^{\infty}$ is L. This means that there exists an integer N, such that for all $n \geq N$, the terms of the sequence are as close to L as we want.

Bounded Sequences. If there exists an $M > 0$ such that $|a_n| \leq M$ for all n, then the sequence is called bounded. Note that this means that there is a number M which is bigger than any term in the sequence and also that $-M$ is smaller than any term in the sequence.

Example 1: Finding Limits of Sequences

Determine whether the following sequence converges or diverges. If it converges, determine the limit.

a) $\left\{\dfrac{5n^2+10}{5n^2-10}\right\}_{n=1}^{\infty}$ b) $\left\{\dfrac{(-1)^n}{n^2+4}\right\}_{n=1}^{\infty}$ c) $\left\{\dfrac{5n^2}{n+1}\right\}_{n=1}^{\infty}$

Strategy. For quotients of polynomials, divide the denominator and the numerator by n to the highest power which appears in the denominator. By doing this all the terms of the form $\dfrac{c}{n^p}, c =$ a constant, will approach 0 as $n \to \infty$. Consequently, all those terms will disappear.

Solution. a) This sequence converges, since

$$\lim_{n\to\infty} \frac{5n^2+10}{5n^2-10} = \lim_{n\to\infty} \frac{5+\frac{10}{n^2}}{5-\frac{10}{n^2}} = \frac{5+0}{5-0} = 1$$

b) Note that $\left|\dfrac{(-1)^n}{n^2+4}\right| = \dfrac{1}{n^2+4}$ and $\lim_{n\to\infty}\dfrac{1}{n^2+4} = 0.$

Corollary 1.1 in the text states: If $\lim_{n\to\infty} |a_n| = 0$, then $\lim_{n\to\infty} |a_n| = 0$, also.

Therefore $\lim_{n\to\infty}\dfrac{(-1)^n}{n^2+4} = 0$. Note that this sequence is an alternating sequence. The terms alternate between being positive and negative.

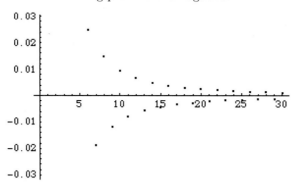

c) This sequence diverges, since

$$\lim_{n\to\infty} \frac{5n^2}{n+1} = \lim_{n\to\infty} \frac{5n}{1+\frac{1}{n}} = \infty$$

Summary. From the above example, we see that the sequence converges if the n-th term approaches a finite number. If the n-th term approaches infinity, then the sequence diverges. Note that this is not the only way to obtain a divergent sequence. We will see other ways as we progress.

Example 2: Using L'Hôpital's Rule to Find a Limit

Evaluate
$$\lim_{n\to\infty} \frac{\ln n + 1}{n}$$

Strategy. When looking at this sequence it is tempting to try to use L'Hôpital's Rule. However, one of the criteria for L'Hôpital's Rule is continuity. Because a sequence is defined only at integers, it cannot be continuous. To solve this problem, we use the corresponding function $f(x) = \dfrac{\ln x + 1}{x}$. We observe that for $x > 0$, this function is continuous. (Note that Theorem 1.2 in the text states that if the $\lim_{x\to\infty} f(x) = L$, then for the corresponding sequence we have $\lim_{n\to\infty} f(n) = L$).

Solution. This limit is of type $\frac{\infty}{\infty}$, therefore we can use L'Hôpital's Rule to find the limit.

$$\lim_{x\to\infty} \frac{\ln x + 1}{x} = \lim_{x\to\infty} \frac{\frac{1}{x}}{1} = 0$$

Therefore
$$\lim_{n\to\infty} \frac{\ln n + 1}{n} = 0$$

Summary. We need to be careful to check the criteria before using L'Hôpital's Rule. Since no sequence can be continuous, we must used the related function.

Example 3: Applying the Squeeze Theorem to A Sequence

Evaluate
$$\lim_{n \to \infty} \frac{\sin n}{2n^3}$$

Strategy. Whenever we are working with the sine or cosine function, we need to think about the fact that both functions are bounded by -1 and 1. We can write
$$-1 \leq \sin n \leq 1$$
$$-1 \leq \cos n \leq 1$$
We often use these inequalities if we want to use the squeeze theorem.

Solution. We know that $\sin n$ is always between -1 and $+1$. Thus

$$-1 < \sin n < +1 \qquad \text{(divide by } 2n^3 > 0\text{)}$$
$$-\frac{1}{2n^3} < \frac{\sin n}{2n^3} < \frac{1}{2n^3}$$

Since $\lim_{n \to \infty} -\frac{1}{2n^3} = 0$, and $\lim_{n \to \infty} \frac{1}{2n^3} = 0$, it follows that the limit of the *squeezed* sequence is also 0.

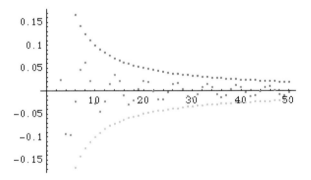

Summary. In this example, we see how we can start with the fact that the sine function is squeezed between -1 and $+1$ to determine the convergence of a function containing $\sin n$.

Example 4: A Monotonic and Bounded Sequence

Investigate whether the sequence $\left\{\dfrac{5^n}{(n+1)!}\right\}_{n=1}^{\infty}$ is increasing, decreasing or neither, and whether the sequence is bounded. Finally, decide whether the sequence is convergent or divergent.

Strategy. In order to decide if a sequence is increasing or decreasing, we can look at $\dfrac{a_{n+1}}{a_n}$ for all $n \geq 1$. If $\dfrac{a_{n+1}}{a_n} \leq 1$, then $a_{n+1} \leq a_n$, and the sequence is decreasing, and if $\dfrac{a_{n+1}}{a_n} \geq 1$, then $a_{n+1} \geq a_n$, and the sequence is increasing.

Solution.

$$\begin{aligned}
\frac{a_{n+1}}{a_n} &= \frac{\frac{5^{n+1}}{((n+1)+1)!}}{\frac{5^n}{(n+1)!}} = \frac{5^{n+1}}{(n+2)!} \cdot \frac{(n+1)!}{5^n} \\
&= \frac{5^{n+1}}{5^n} \cdot \frac{(n+1)!}{(n+2)!} = \frac{5^n \cdot 5}{5^n} \cdot \frac{(n+1)!}{(n+2)(n+1)!} \\
&= \frac{5}{n+2} \leq 1 \text{ for } n \geq 3
\end{aligned}$$

$$\implies \frac{a_{n+1}}{a_n} \leq 1 \text{ for } n \geq 3$$
$$\implies a_{n+1} \leq a_n \text{ for } n \geq 3$$

Therefore the sequence is decreasing for $n \geq 3$.

Is it bounded from above? Yes, since $a_{n+1} \leq a_n$ for $n \geq 3$, it follows that

$$a_n \leq a_3 = \frac{5^3}{4!} = \frac{125}{24} \text{ for } n \geq 3$$

$a_1 = \frac{5}{2}$, $a_2 = \frac{25}{6}$. Since all of these terms are less than 5, it follows that the sequence is bounded from above by 5.

Also, since all of the terms are positive, it is bounded below by 0.

Since the sequence is decreasing and bounded, it follows that it converges.

Summary. It was easy to show that this sequence was bounded below since all the terms are positive. We defined bounded by saying that there exists an $M > 0$ such that $|a_n| \leq M$ for all n. In this problem, $M = 5$, which says that $-5 \leq a_n \leq 5$ which is a true statement. If it is bounded below by 0, then it is certainly bounded below by -5, also.

7.2. Infinite Series

Key Topics.
- Series
- Geometric Series
- k-th Term Test for Divergence

Worked Examples.
- A Telescoping Series
- Examples of Geometric Series
- Series Whose Terms Do Not Tend to Zero

Overview

This section begins our study of series which we will continue throughout this chapter. We look at the first of many theorems concerning convergence or divergence of a series.

Series. If we take an infinite sequence, we can add the first two terms and call the sum S_2, or the first three terms and call the sum S_3, or, in general, add up the first n terms and call the sum S_n. Each of these sums is called a partial sum. We can make a sequence of all the answers to the n-th partial sums, beginning with $S_1 = a_1$.

When we add up these terms we write $\sum_{k=1}^{\infty} a_k$. This is called an infinite **series**. Note that the difference between a sequence and a series is that a sequence is a list of numbers series is the sum of those numbers. If the sequence of partial sums $S_n = \sum_{k=1}^{\infty} a_k$ converges to a finite number S, then the series $\sum_{k=1}^{\infty} a_k$ converges to S also.

Geometric Series. A series is a geometric series if we go from one term to the next by multiplying by a number r. r is called the common ratio and the series can be written as:
$$a + ar + ar^2 + ar^3 + ar^4 + ...$$
More compactly, $\sum_{k=1}^{\infty} ar^k$ is a geometric series. This series converges to $S = \dfrac{a}{1-r}$ if $|r| < 1$. Otherwise, it diverges.

k-th Term Test for Divergence. In a series, if $\lim_{k \to \infty} a_k \neq 0$, then the series diverges.

Careful! The only conclusion that can be made using this test is that a series diverges. If $\lim_{k \to \infty} a_k = 0$, then there is NO CONCLUSION! Another test must be used.

Example 1: A Telescoping Series

Investigate the convergence or divergence of the series

$$\sum_{k=1}^{\infty} \ln\left(\frac{k+1}{k}\right)$$

Strategy. We need to start by rewriting the problem using one of the properties of logarithms.

$$\ln\left(\frac{k+1}{k}\right) = \ln(k+1) - \ln(k)$$

Solution. Consider the n-th partial sum

$$
\begin{aligned}
S_n &= \sum_{k=1}^{n} (\ln(k+1) - \ln(k)) \\
&= (\ln 2 - \ln 1) + (\ln 3 - \ln 2) + (\ln 4 - \ln 3) + \ldots + (\ln n - \ln(n-1)) + (\ln(n+1) - \ln n) \\
&= -\ln 1 + \ln(n+1) = \ln(n+1)
\end{aligned}
$$

Thus $S_n = \ln(n+1)$, and therefore $\lim_{n\to\infty} S_n = \lim_{n\to\infty} \ln(n+1) = \infty$. Thus the series diverges.

Summary. This type of series is called telescoping because the middle terms all cancel out and the series collapses or *telescopes*.

Example 2: Examples of Geometric Series

Investigate the convergence of the following geometric series.

a) $\sum_{k=0}^{\infty} 5 \cdot \left(\frac{2}{3}\right)^k$

b) $\sum_{k=0}^{\infty} \frac{3}{4^{k-1}}$

c) $\sum_{k=2}^{\infty} \frac{3+(-3)^k}{5^k}$

d) $\sum_{k=0}^{\infty} \frac{4^{k+1}}{3^k}$

e) $\frac{9}{10} + \frac{9}{10^2} + \frac{9}{10^3} + \ldots$

Strategy. For each example, we will first determine if it is a geometric series. In other words, we will see if there is a common ratio, r. If there is, then if $|r| < 1$ we conclude that the series converges to $S = \frac{a}{1-r}$ where a is the first term. If $|r| \geq 1$ then the series diverges.

Solution. a) $\sum_{k=0}^{\infty} 5 \cdot \left(\frac{2}{3}\right)^k = 5 \cdot \left(\frac{2}{3}\right)^0 + 5 \cdot \left(\frac{2}{3}\right)^1 + 5 \cdot \left(\frac{2}{3}\right)^2 + 5 \cdot \left(\frac{2}{3}\right)^3 + \ldots$

We see that $r = \frac{2}{3} < 1$ and hence the series converges. We have $a = 5$, $r = \frac{2}{3}$, and thus $\sum_{k=0}^{\infty} 5 \cdot \left(\frac{2}{3}\right)^k = \frac{5}{1-\frac{2}{3}} = \frac{5}{\frac{1}{3}} = 5 \cdot \frac{3}{1} = 15$

b) $\sum_{k=0}^{\infty} \frac{3}{4^{k-1}} = \frac{3}{4^{-1}} + \frac{3}{4^0} + \frac{3}{4^1} + \frac{3}{4^2} + \frac{3}{4^3} + \ldots$

By writing out several terms, it helps us to see the pattern and then we can rewrite the series as $\sum_{k=0}^{\infty} \frac{12}{4^k}$ if we want.

Now we can see that the first term is $a = 12$, and the common ratio is $r = \frac{1}{4}$. The sum of the series is:

$$S = \frac{a}{1-r} = \frac{12}{1-\frac{1}{4}} = \frac{12}{\frac{3}{4}} = 12 \cdot \frac{4}{3} = 16$$

c) The series $\sum_{k=2}^{\infty} \dfrac{3+(-3)^k}{5^k}$ can be written as two series that we add together.

$$\sum_{k=2}^{\infty} \frac{3+(-3)^k}{5^k} = \sum_{k=2}^{\infty} \frac{3}{5^k} + \sum_{k=2}^{\infty} \frac{(-3)^k}{5^k} = \sum_{k=2}^{\infty} 3\left(\frac{1}{5}\right)^k + \sum_{k=2}^{\infty} \left(\frac{-3}{5}\right)^k$$

For the first series, $\sum_{k=2}^{\infty} 3\left(\frac{1}{5}\right)^k$, the first term is $a = \frac{3}{25}$, and the ratio is $r = \frac{1}{5}$.

Simmilarly in the second series $\sum_{k=2}^{\infty} \left(\frac{-3}{5}\right)^k$, $a = \frac{9}{25}$, and $r = \frac{-3}{5}$, thus we get

$$\sum_{k=2}^{\infty} \frac{3+(-3)^k}{5^k} = \frac{\frac{3}{25}}{1-\frac{1}{5}} + \frac{\frac{9}{25}}{1-\left(\frac{-3}{5}\right)} = \frac{3}{25} \cdot \frac{5}{4} + \frac{9}{25} \cdot \frac{5}{8} = \frac{3}{20} + \frac{9}{40} = \frac{3}{8}$$

d) $\sum_{k=0}^{\infty} \dfrac{4^{k+1}}{3^k} = \sum_{k=0}^{\infty} 4 \cdot \dfrac{4^k}{3^k} = \sum_{k=0}^{\infty} 4\left(\dfrac{4}{3}\right)^k$ diverges, since the ratio $r = \dfrac{4}{3} > 1$.

e) The first term of this geometric series is $a = \frac{9}{10}$, and the ratio $r = \frac{1}{10}$, thus

$$S = \frac{\frac{9}{10}}{1-\frac{1}{10}} = \frac{\frac{9}{10}}{\frac{9}{10}} = 1$$

Note that we can also rewrite this series as

$$\frac{9}{10} + \frac{9}{10^2} + \frac{9}{10^3} + \ldots = \sum_{k=0}^{\infty} 9\left(\frac{1}{10}\right)^{k+1}$$

Summary. It really helps to write out several terms of the series so that you can see what it looks like before deciding on a way to test for convergence.

Example 3: Series Whose Terms Do Not Tend to Zero

Investigate the series for convergence.

$$\text{a) } \sum_{k=1}^{\infty} \frac{2k+1}{2k-1} \text{ and b) } \sum_{k=1}^{\infty} \cos(k\pi)$$

Strategy. The very first test to perform when looking at convergence of series is the k-th test for divergence. If the k-th term does not approach zero, then the series diverges and the problem is complete!

Solution. a) $\sum_{k=1}^{\infty} \frac{2k+1}{2k-1}$

Look at $\lim\limits_{k\to\infty} \frac{2k+1}{2k-1} = \frac{2}{2} = 1 \neq 0$ and the series diverges.

b) $\cos(k\pi) = \begin{cases} +1 \text{ for } k \text{ even} \\ -1 \text{ for } k \text{ odd} \end{cases}$

Thus the terms of the sequence $\{\cos(k\pi)\}_{k=1}^{\infty}$ alternates between $+1$ and -1, and in particular $\lim\limits_{k\to\infty} \cos(k\pi) \neq 0$, therefore $\sum_{k=1}^{\infty} \cos(k\pi)$ diverges, also.

Summary. As the chapter continues, we are going to learn about other tests for convergence and divergence. It is a good idea to always start with the k-th term test for divergence, so that if $\lim\limits_{k\to\infty} a_k \neq 0$, we can conclude that the series diverges and go on to the next problem.

7.3. The Integral Test and Comparison Tests

Key Topics.
- The Integral Test
- p-Series
- Direct Comparison Test
- Limit Comparison Test

Worked Examples.
- Using the Integral Test
- p-Series
- Using the Comparison Tests
- Using the Comparison Test for a Divergent Series

Overview

Obviously and unfortunately, all series are not geometric series. In this section we study several other test we can use to test for convergence or divergence of a series.

The Integral Test. If $f(x) = a_k$ for each non-negative integer k, and $f(x)$ is a function which is continuous, and decreasing, and $f(x) \geq 0$ for $x \geq 1$ then $\int_1^\infty f(x)dx$ and the series $\sum_{k=1}^\infty a_k$ either both converge or both diverge. In other words, if the function which corresponds to the series is continuous, decreasing, and positive, then the improper integral of that function integrated from 1 to ∞ and the infinite series act the same way. They either both converge or both diverge.

p-Series. The p-series is really nice, because we can tell if the series converges just by looking at it. $\sum_{k=1}^\infty \frac{1}{k^p}$ converges for $p < 1$ and diverges for $p \geq 1$. This is easily proved by the integral test.

Direct Comparison Test. If we have a series $\sum_{k=1}^\infty a_k$, then if we can find a **convergent** series $\sum_{k=1}^\infty b_k$, and $0 \leq a_k \leq b_k$, then $\sum_{k=1}^\infty a_k$ converges also.

Think about this - if the larger series converges the smaller one must also. The smaller one cannot go off to ∞ and still remain smaller than the convergent series.

Still looking at the above two series, with $0 \leq a_k \leq b_k$, if $\sum_{k=1}^{\infty} a_k$ diverges, then so does $\sum_{k=1}^{\infty} b_k$.

Think about this one. The smaller series is going off to infinity. Therefore, the larger one must go off to infinity also or it wouldn't be larger for all k.

It is extremely important to understand which way the inequality must go in order to use this test.

Limit Comparison Test. This test is a little easier to use than the direct comparison test. In this test we find the series we want to use to compare to the one in question, and then take the limit of their ratio.

If $a_k > 0$, and $b_k > 0$ and if for some finite positive value L, $\lim_{k \to \infty} \dfrac{a_k}{b_k} = L$, then either $\sum_{k=1}^{\infty} a_k$ and $\sum_{k=1}^{\infty} b_k$ either both converge or both diverge.

We often use the p-series as the comparison series since we easily know whether that converges or diverges.

Example 1: Using the Integral Test

Investigate the convergence or divergence of the series

$$\sum_{n=2}^{\infty} \frac{1}{n \ln n}$$

Strategy. The function corresponding to this series is $f(x) = \dfrac{1}{x \ln x}$. This is a continuous, non-negative function on the interval $[2, \infty)$. We need to show that it is decreasing in order to use the integral test. We will do this by finding the derivative $f'(x)$ and seeing that it is less than zero for all x, thus telling us that $f(x)$ is a decreasing function.

Solution. We use the quotient rule to find the derivative $f'(x)$.

$$f'(x) = \frac{0 - (\ln x + x \cdot \frac{1}{x})}{(x \ln x)^2} = \frac{-\ln x - 1}{(x \ln x)^2} < 0 \text{ for } x \geq 2$$

Therefore f is decreasing for $x \geq 2$ and we can use the integral test.

$$\int_2^{\infty} \frac{dx}{x \ln x} \qquad \text{(rewrite as a limit)}$$

$$= \lim_{b \to \infty} \int_2^b \frac{dx}{x \ln x} \qquad \text{(change limits of integration)}$$

$$= \lim_{b \to \infty} \int_{\ln 2}^{\ln b} \frac{du}{u}$$

$$= \lim_{b \to \infty} \ln |u| \, \Big|_{\ln 2}^{\ln b}$$

$$= \lim_{b \to \infty} (\ln(\ln b) - \ln(\ln 2))$$

$$= \infty$$

Therefore the series $\sum_{n=2}^{\infty} \dfrac{1}{n \ln n}$ diverges.

Summary. Don't forget to check the criteria for the use of the integral test before using it.

Example 2: p-Series

Investigate the convergence or divergence of the following series.

$$\text{a) } \sum_{n=2}^{\infty} \frac{2}{\sqrt{n^3}} \text{ and b) } \sum_{n=1}^{\infty} 2n^{-\frac{1}{2}}$$

Strategy. For the p-series, we have to write the series so that it is in the form $\sum \frac{1}{n^p}$ first.

Solution. a) $\sum_{n=2}^{\infty} \frac{2}{\sqrt{n^3}} = 2 \sum_{n=2}^{\infty} \frac{1}{n^{\frac{3}{2}}}$ which show us that $p = \frac{3}{2} > 1$ and the series converges.

b) But $\sum_{n=1}^{\infty} 2n^{-\frac{1}{2}} = 2 \sum_{n=2}^{\infty} \frac{1}{n^{\frac{1}{2}}}$ diverges, since $p = \frac{1}{2} < 1$.

Summary. If you remember that the form of the p-series must be of the form $\sum_{n=c}^{\infty} \frac{1}{n^p}$, where c is a constant, then this test should be one of your favorites as there is very little work required to obtain the conclusion.

Example 3: Using the Comparison Tests

Investigate the convergence or divergence of the following series.

$$\text{a) } \sum_{k=1}^{\infty} \frac{10}{k^2 + 10} \text{ and b) } \sum_{k=1}^{\infty} \frac{10}{k^2 - 10}$$

Strategy. In both cases we will use $\sum_{k=1}^{\infty} \frac{10}{k^2}$ as our series for comparison. In part a) we can use either direct comparison or limit comparison but, as you will see, in part b) the limit comparison test will be easier to use, since it is difficult to find an inequality that works for direct comparison.

Solution. a) $\sum_{k=1}^{\infty} \frac{10}{k^2 + 10}$

We can use the direct comparison test for this series, since it is easy to find an inequality:

$$a_k \leq \frac{10}{k^2 + 10} \leq \frac{10}{k^2}$$

Since $\sum_{k=1}^{\infty} \frac{10}{k^2}$ converges (p-series, $p = 2$), so does $\sum_{k=1}^{\infty} \frac{10}{k^2 + 10}$ by direct comparison.

b) Unfortunately, in this case $\sum_{k=1}^{\infty} \frac{10}{k^2 - 10}$, we have $a_k = \frac{10}{k^2 - 10} > \frac{10}{k^2}$

Since the convergent p-series $\sum_{k=1}^{\infty} \frac{10}{k^2}$ is smaller than the given series, no conclusion can be formed. Instead, then, let's try the Limit Comparison Test.

We will compare $\sum_{k=1}^{\infty} \frac{10}{k^2 - 10}$ to the convergent series $\sum_{k=1}^{\infty} \frac{10}{k^2}$. If $a_k = \frac{10}{k^2 - 10}$, and $b_k = \frac{10}{k^2}$, then

$$\lim_{k \to \infty} \frac{a_k}{b_k} = \lim_{k \to \infty} \frac{10}{k^2 - 10} \cdot \frac{k^2}{10} = \lim_{k \to \infty} \frac{10k^2}{10k^2 - 100} = \lim_{k \to \infty} \frac{10}{10 - \frac{100}{k^2}} = 1$$

Since this limit, L, is a finite positive number, either both series converge or both series diverge. We know that $\sum_{k=1}^{\infty} \frac{10}{k^2}$ is a convergent p-series, and thus $\sum_{k=1}^{\infty} \frac{10}{k^2 - 10}$ converges also.

Summary. Try to **really** understand the inequality in the Direct Comparison Test. In that way, you will use the test correctly. However, as stated earlier, the Limit Comparison Test is often an easier test to use. We could have used it for part a) in this problem, also.

Example 4: Using the Comparison Test for a Divergent Series

Investigate the convergence or divergence of the series $\sum_{k=1}^{\infty} \frac{2^k}{2k-1}$.

Strategy. We will compare this series to the divergent harmonic series $\sum_{k=1}^{\infty} \frac{1}{2k}$.

Solution. Since
$$\frac{2^k}{2k-1} \geq \frac{1}{2k} \geq 0$$
and since $\sum_{k=1}^{\infty} \frac{1}{2k} = \frac{1}{2} \sum_{k=1}^{\infty} \frac{1}{k}$ diverges, so does $\sum_{k=1}^{\infty} \frac{2^k}{2k-1}$.

Summary. Remember that in a fraction as the top gets bigger, the fraction gets bigger. But, as the bottom gets bigger, the fraction gets smaller. In this example, 2^k grows faster than $2k-1$, thus $\lim \frac{2^k}{2k-1} = \infty \neq 0$. Thus, we could have used the k-th term test, also, to show that this series diverges.

7.4. Alternating Series

Key Topics.
- Alternating Series Test
- Estimating the Sum of an Alternating Series

Worked Examples.
- Using the Alternating Series Test
- Estimating the Sum of an Alternating Series

Overview

In this section we discuss series that have a particular pattern, in that the terms alternate from positive to negative (or negative to positive). Note that $(-1)^k$ or $(-1)^{k+1}$ are both sequences alternating between $+1$, and -1. An alternating series can be written as

$$\sum_{k=1}^{\infty}(-1)^{k+1}a_k = a_1 - a_2 + a_3 - a_4 + ...$$

$$a_k \geq 0 \text{ for all } k \geq 1$$

The Alternating Series Test. An alternating series will converge if both of the following two conditions hold:

i) $a_{k+1} \leq a_k$ (sequence $\{a_k\}$ is decreasing)

ii) $\lim_{k \to \infty} a_k = 0$

This tells us that the terms of the series are decreasing and that the limit of the k-th term is zero.

Do NOT confuse this last statement with the k-th series test for divergence. Only for an alternating series can we use the limit of the k-th term to show convergence. A common mistake is to assume that a series that is not alternating converges if the limit of the k-th term is zero.

If $\lim_{k \to \infty} a_k = 0$, then $\sum_{k=1}^{\infty} a_k$ might, or might not converge.

However, if i) and ii) above are satisfied for an alternating series, then the alternating series $\sum_{k=1}^{\infty}(-1)^{k+1}a_k$ does converge.

Estimating the Sum of an Alternating Series. For an alternating series, we can find the n-th partial sum, S_n, for any n we choose, just by adding up the first n terms. We like to know how close our estimate is to the actual sum ,S, which we cannot physically find. The theorem tells us that $|S - S_n| \leq a_{n+1}$. In other words, the error is smaller than or equal to the absolute value of the first *neglected* term. If, for example, we have found S_6, then the error is not greater than a_7.

Example 1: Using the Alternating Series Test

Determine if the series $\sum_{k=1}^{\infty} \frac{(-1)^{k+1}}{2k-1}$ converges or diverges.

Strategy. We can write out the first few terms to see what the series looks like. We need to show that the absolute value of the terms is decreasing, and that $\lim_{k\to\infty} a_k = 0$. If those two conditions are met, we can conclude that the alternating series converges.

Solution. First, we check if the sequence is decreasing.

$$a_k = \frac{1}{2k-1}$$
$$a_{k+1} = \frac{1}{2(k+1)-1} = \frac{1}{2k+2-1} = \frac{1}{2k+1}$$
$$\implies a_{k+1} = \frac{1}{2k+1} < \frac{1}{2k-1} = a_k$$

Thus, the terms are decreasing.

Next, we examine the limit of a_k.

$$\lim_{k\to\infty} a_k = \lim_{k\to\infty} \frac{1}{2k-1} = 0$$

Since both conditions are met, the series $\sum_{k=1}^{\infty} \frac{(-1)^{k+1}}{2k-1}$ converges by the alternating series test.

Summary. Make sure to check both conditions for the alternating series test.

Example 2: Estimating the Sum of an Alternating Series

Approximate the sum $\sum_{k=1}^{\infty} \frac{(-1)^{k+1}}{2k-1}$ using the 4-th partial sum and estimate the error in this approximation.

Strategy. We will find S_4 by adding the first four terms of the series. We know that the error will not be greater than a_5.

Solution. We know from the previous example that this series converges.

$$\begin{aligned} S_4 &= \frac{1}{2-1} - \frac{1}{4-1} + \frac{1}{6-1} - \frac{1}{8-1} \\ &= \frac{1}{1} - \frac{1}{3} + \frac{1}{5} - \frac{1}{7} \\ &= \frac{105 - 35 + 21 - 15}{105} = \frac{76}{105} \approx 0.7238 \end{aligned}$$

The error in this approximation is smaller than $a_5 = \frac{1}{10-1} = \frac{1}{9} \approx .111$.

Summary. By adding only 4 terms of this infinite series, we end up with a sum which differs from the actual sum by at most $\frac{1}{9}$. Adding the terms is not difficult. We could get as close as we wish to the actual sum by adding a finite number of terms.

7.5. Absolute Convergence And The Ratio Test

Key Topics.
- Absolute Convergence
- The Ratio Test
- The Root Test

Worked Examples.
- Testing for Absolute Convergence
- Using the Ratio Test
- Using the Root Test

Overview

We saw in section 7.4 that part of the alternating series test involves finding the limit of the k-th term. As pointed out, we can find convergence using the limit of the k-th term **only** if we have an alternating series. What about other series that have positive and negative terms? In this section, we look at the series of absolute values, that is for $\sum\limits_{k=1}^{\infty} a_k$, we will investigate the series $\sum\limits_{k=1}^{\infty} |a_k|$ for convergence. Since all terms of the series $\sum\limits_{k=1}^{\infty} |a_k|$ are positive, we can use the tests we previously learned.

Absolute Convergence. If the series of absolute values converges, then we say that the original series **converges absolutely**. Note that absolute convergence means that the series of absolute values converges. That is why it is called *absolutely* convergent. Note that if the series converges absolutely, we can also say that the series converges.

$$\text{If } \sum_{k=1}^{\infty} |a_k| \text{ converges, so does } \sum_{k=1}^{\infty} a_k \ .$$

There are many series that are convergent, but not absolutely convergent These are called **conditionally convergent**. The alternating harmonic series, $\sum\limits_{k=1}^{\infty} \frac{(-1)^{k+1}}{k}$, is an example of a conditionally convergent series. The alternating harmonic series converges, but since $\sum\limits_{k=1}^{\infty} \frac{1}{k}$, the harmonic series, diverges, the series $\sum\limits_{k=1}^{\infty} \frac{(-1)^{k+1}}{k}$ does not converge absolutely.

The Ratio Test. If we are looking for convergence of a series containing factorials and/or exponentials, the ratio test works wonderfully. It tells us to look at the limit of the ratio $\left|\frac{a_{k+1}}{a_k}\right|$.

$$\lim_{k \to \infty} \left|\frac{a_{k+1}}{a_k}\right| = L$$

If $L < 1$, the series converges absolutely.

If $L > 1$, the series diverges.

If $L = 1$, there is no conclusion; use another test.

The Root Test. The root test has a similar conclusion to the ratio test. We use the k-th root of the absolute value of the k-th term of the series to determine convergence.

$$\lim_{k \to \infty} \sqrt[k]{|a_k|} = L$$

If $L < 1$, the series converges absolutely.

If $L > 1$, the series diverges.

If $L = 1$, there is no conclusion; use another test.

Example 1: Testing for Absolute Convergence

a) Does the series $\sum_{k=1}^{\infty} \frac{(-1)^{k+1}}{k^2}$ converge absolutely?

b) Does the series $\sum_{k=1}^{\infty} \frac{(-1)^{k+1}}{k^2}$ converge?

Strategy. We will check for absolute convergence by replacing the $(-1)^{k+1}$ with 1. In that way we have a series of positive terms. If the series converges absolutely, then we can conclude that the series converges.

Solution.
$$\sum_{k=1}^{\infty} \left| \frac{(-1)^{k+1}}{k^2} \right| = \sum_{k=1}^{\infty} \frac{1}{k^2} \text{ converges } (p\text{-series}, p = 2 > 1)$$

Thus $\sum_{k=1}^{\infty} \frac{(-1)^{k+1}}{k^2}$ converges absolutely, and, therefore, $\sum_{k=1}^{\infty} \frac{(-1)^{k+1}}{k^2}$ converges.

Summary. To show absolute convergence we can use any of the tests we have studied so far. In this problem, we had a convergent p-series. Absolute convergence implies convergence, and we are finished.

Example 2: Using the Ratio Test

a) Use the ratio test to test $\sum_{k=1}^{\infty} \dfrac{5^k}{k!}$ for convergence.

b) Does $\sum_{k=1}^{\infty} \dfrac{(-5)^k}{k!}$ converge, also?

Strategy. We will use the ratio test here.

Solution. a)

$$\lim_{k\to\infty} \left|\frac{a_{k+1}}{a_k}\right| = \lim_{k\to\infty} \frac{5^{k+1}}{(k+1)!}\frac{k!}{5^k}$$
$$= \lim_{k\to\infty} \frac{5}{(k+1)} = 0 < 1$$

Thus, since $L < 1$, the series $\sum_{k=1}^{\infty} \dfrac{5^k}{k!}$ converges absolutely by the ratio test.

b) Since $\sum_{k=1}^{\infty} \left|\dfrac{(-5)^k}{k!}\right| = \sum_{k=1}^{\infty} \dfrac{5^k}{k!}$ converges absolutely, which is a stronger test than conditional convergence, we know that the alternating series $\sum_{k=1}^{\infty} \dfrac{(-5)^k}{k!}$ also converges (Absolute Convergence Test).

Summary. Whenever possible, try an absolute convergence test first. Then, if that doesn't work and you have an alternating series, use the alternating series test and test for conditional convergence.

Example 3: Using the Root Test

Determine whether the following series are convergent.

$$\text{a) } \sum_{k=1}^{\infty} \left(\frac{k}{2k+1}\right)^k$$

$$\text{b) } \sum_{k=1}^{\infty} \left(\frac{2k+1}{k}\right)^k$$

$$\text{c) } \sum_{k=1}^{\infty} \left(\frac{k+1}{k}\right)^k$$

Strategy. Since each of these series include exponentials, we will use the root test.

Solution. a) $\sum_{k=1}^{\infty} \left(\frac{k}{2k+1}\right)^k$

$$\lim_{k\to\infty} \sqrt[k]{\left|\left(\frac{k}{2k+1}\right)^k\right|} = \lim_{k\to\infty} \frac{k}{2k+1} = \frac{1}{2} < 1$$

$L < 1$ implies that this series converges.

b) $\sum_{k=1}^{\infty} \left(\frac{2k+1}{k}\right)^k$

$$\lim_{k\to\infty} \sqrt[k]{\left|\left(\frac{2k+1}{k}\right)^k\right|} = \lim_{k\to\infty} \frac{2k+1}{k} = 2 > 1$$

$L > 1$ implies that this series diverges.

c) $\sum_{k=1}^{\infty} \left(\frac{k+1}{k}\right)^k$

$$\lim_{k\to\infty} \sqrt[k]{\left|\left(\frac{k+1}{k}\right)^k\right|} = \lim_{k\to\infty} \frac{k+1}{k} = 1$$

Since $L = 1$, the ratio test is inconclusive.
Let's use the k-th term test for divergence.

$$\lim_{k\to\infty} \left(\frac{k+1}{k}\right)^k = \left(1 + \frac{1}{k}\right)^k = e \neq 0$$

It follows that $\sum_{k=1}^{\infty} \left(\frac{k+1}{k}\right)^k$ diverges, since the terms do not approach zero.

Summary. The series in parts a) and b) are reciprocals of each other. Note how that change causes one series to converge and the other to diverge.

In part c) we couldn't conclude anything from the root test, and therefore had to choose a different test.

7.6. Power Series

Key Topics.
- Power Series
- Radius of Convergence
- Differentiating and Integrating Power Series

Worked Examples.
- A Power Series That Converges Everywhere
- Finding the Interval and Radius of Convergence
- Differentiating and Integrating a Power Series

Overview

If you've been questioning why we've been looking at convergent series for so many sections, the rest of this chapter should help answer that question. We begin with power series, in which the terms of the series are functions of x. Note that there have not been x-variables in any of the series so far - we have only used the variable k. We are now adding another variable, x.

Power Series. A power series is a series of the form:

$$\sum_{k=0}^{\infty} b_k(x-c)^k = b_0 + b_1(x-c) + b_2(x-c)^2 + b_3(x-c)^3 + \ldots$$

b_k, $k = 0, 1, 2\ldots$ are called the **coefficients** of the series.

Note that if the series converges, then this is a function of x. Our job is to determine the values of x, which will make the series converge. The ratio and root test are the tests we will use most often to test convergence of power series.

Radius of Convergence. If we use the ratio or root test for power series, then the limit of $\left|\frac{a_{k+1}}{a_k}\right|$ for the ratio test, or the limit of the k-th root, in the case of the root test, will be a function of x, instead of a finite number L as it was before. If we call this function $f(x)$, then the series converges when this limit is less than 1. We need to solve the inequality $|f(x)| < 1$ to find the values of x for which the series converges.

The answer will be an interval, called the **interval of convergence**. Half of the length of the entire interval is called the **radius of convergence**. For example, if the series converges on the interval $(-5, 5)$ then the radius of convergence is $r = 5$, since the length of the interval from -5 to 5 is 10.

Differentiating and Integrating Power Series. Power series can be differentiated and integrated term by term. In both cases the radius of convergence, r, remains the same. Note that this statement is true only for power series. Other series must be checked individually.

Example 1: A Power Series That Converges Everywhere

Determine the values of x for which the power series $\sum_{k=0}^{\infty} \frac{x^k}{k!}$ converges.

Strategy. As we usually do with power series, we use the ratio test to determine the values of x for which this series converges.

Solution. From the ratio test, we have

$$\lim_{k\to\infty} \left|\frac{a_{k+1}}{a_k}\right| = \lim_{k\to\infty} \left|\frac{x^{k+1}}{(k+1)!} \cdot \frac{k!}{x^k}\right| = \lim_{k\to\infty} \left|\frac{x}{k+1}\right| = 0 < 1$$

Therefore $\sum_{k=0}^{\infty} \frac{x^k}{k!}$ converges for all x. In this case the radius of convergence $r = \infty$.

Summary. This series is a very special series as we will see in the next section. It is the Taylor series of e^x, and in particular $\sum_{k=0}^{\infty} \frac{x^k}{k!} = e^x$ for all x.

Example 2: Finding the Interval and Radius of Convergence

Determine the interval and radius of convergence for the power series

$$\sum_{k=1}^{\infty} \frac{(x-1)^k}{3^k k}$$

Strategy. We will use the ratio test. The limit of $\left|\frac{a_{k+1}}{a_k}\right|$ will end up as a function of x. We will then solve the inequality $|f(x)| < 1$ to find the interval and the radius of convergence. We then need to test each endpoint of the interval separately to determine if we have an open, closed, or half-open interval of convergence.

Solution. From the ratio test, we have

$$\lim_{k \to \infty} \left|\frac{a_{k+1}}{a_k}\right| = \lim_{k \to \infty} \left|\frac{(x-1)^{k+1}}{3^{k+1}(k+1)} \cdot \frac{3^k k}{(x-1)^k}\right|$$

$$= \lim_{k \to \infty} \left|\frac{x-1}{3} \cdot \frac{k}{k+1}\right|$$

$$= \left|\frac{x-1}{3}\right| \lim_{k \to \infty} \frac{k}{k+1} = \left|\frac{x-1}{3}\right|$$

Thus the series $\sum_{k=1}^{\infty} \frac{(x-1)^k}{3^k k}$ converges for

$$\left|\frac{x-1}{3}\right| < 1 \implies |x-1| < 3 \implies \text{radius of convergence is 3}$$

$$-3 < x - 1 < 3$$
$$-2 < x < 4$$

At this point we know that the series converges absolutely for $-2 < x < 4$, and diverges for $x < -2$, and $x > 4$.

We need to test the series for convergence at the endpoints $x = -2$ and $x = 4$.

When $x = 4$, we get

$$\sum_{k=1}^{\infty} \frac{(4-1)^k}{3^k k} = \sum_{k=1}^{\infty} \frac{3^k}{3^k k} = \sum_{k=1}^{\infty} \frac{1}{k}$$

This is the harmonic series, and thus diverges.

When $x = -2$, we get

$$\sum_{k=1}^{\infty} \frac{(-2-1)^k}{3^k k} = \sum_{k=1}^{\infty} \frac{(-3)^k}{3^k k} = \sum_{k=1}^{\infty} \frac{(-1)^k}{k}$$

This is the alternating harmonic series, and thus converges.

The interval of convergence for the power series $\sum_{k=1}^{\infty} \frac{(x-1)^k}{3^k k}$ is thus $[-2, 4)$, and the radius of convergence is 3.

Summary. There are many steps in this process. Use the knowledge of the convergence of a power series to find the interval of convergence. After all that, we need to test each endpoint separately.

Example 3: Differentiating and Integrating a Power Series

Differentiate, and integrate the power series $P(x) = \sum_{k=0}^{\infty} \dfrac{x^k}{k!}$.

Strategy. We know from our first example that $P(x)$ converges for all x. Thus we can differentiate and integrate the series $P(x)$ by differentiating and integrating each term. Note that the factorials $3!$, $4!$, $5!$ etc. are all constants when differentiating with respect to x.

Solution.

$$P(x) = \sum_{k=0}^{\infty} \frac{x^k}{k!} = 1 + x + \frac{x^2}{2} + \frac{x^3}{3!} + \frac{x^4}{4!} + \frac{x^5}{5!} + \ldots$$

$$P'(x) = 0 + 1 + \frac{2x}{2} + \frac{3x^2}{3!} + \frac{4x^3}{4!} + \frac{5x^4}{5!} + \ldots$$

$$\implies P'(x) = \sum_{k=1}^{\infty} \frac{kx^{k-1}}{k!}$$

Note that we can rewrite this series.

$$\sum_{k=1}^{\infty} \frac{kx^{k-1}}{k!} = \sum_{k=1}^{\infty} \frac{x^{k-1}}{(k-1)!} = \sum_{k=0}^{\infty} \frac{x^k}{k!} = P(x)$$

($P'(x) = P(x)$ since this series is equal to e^x as we will see in the next section)

$$\int P(x)dx = \int \left(1 + x + \frac{x^2}{2} + \frac{x^3}{3!} + \frac{x^4}{4!} + \frac{x^5}{5!} + \ldots\right) dx$$

$$= x + \frac{x^2}{2} + \frac{x^3}{3 \cdot 2} + \frac{x^4}{4 \cdot 3!} + \frac{x^5}{5 \cdot 4!} + \frac{x^6}{6 \cdot 5!} + \ldots + c$$

$$= x + \frac{x^2}{2} + \frac{x^3}{3!} + \frac{x^4}{4!} + \frac{x^5}{5!} + \ldots + c$$

$$= \sum_{k=1}^{\infty} \frac{x^k}{k!} + c$$

Summary. This again is almost the same as $\sum_{k=0}^{\infty} \dfrac{x^k}{k!}$. What is different is the starting point, so that here we have $a_0 = \dfrac{x^0}{0!} = 1$, a constant. If you think of $P(x) = e^x$, then this shows that $\int P(x)dx = \sum_{k=1}^{\infty} \dfrac{x^k}{k!} + c = P(x) - a_0 + c = P(x) + C$, where $C = c - 1$ is another constant. In other words, $\int e^x dx = e^x + C$.

7.7. Taylor Series

Key Topics.
- Taylor Polynomials and Series
- Error Approximations

Worked Examples.
- Finding a Taylor Series
- Estimating the Error in a Taylor Polynomial Approximation
- Finding New Taylor Series from Old

Overview

We will see in this section that we can use series to help us to work with transcendental functions. We will write transcendental functions as power series. We can think of power series as *infinite* polynomials, and it is often easier to work with polynomials than with transcendental functions.

Taylor Polynomials and Series. If we start with an infinitely differentiable function, then we can construct the series

$$\sum_{k=0}^{\infty} \frac{f^{(k)}(c)}{k!}(x-c)^k$$

This is called a Taylor series expansion for the function f about c. If $c = 0$, we have a special case of a Taylor series called a *MacLaurin* series.

A Taylor polynomial of degree n is written as

$$\begin{aligned} P_n(x) &= \sum_{k=0}^{n} \frac{f^{(k)}(c)}{k!}(x-c)^k \\ &= f(c) + f'(c)(x-c) + \frac{f''(c)}{2!}(x-c)^2 + \ldots + \frac{f^{(n)}(c)}{n!}(x-c)^n \end{aligned}$$

Error Approximations. The error in using the approximation $P_n(x)$ is notated by $R_n(x)$, called the remainder term, and defined as follows.

$$\begin{aligned} R_n(x) &= f(x) - P_n(x) \\ &= \frac{f^{(n+1)}(z)}{(n+1)!}(x-c)^{n+1} \text{ for some } z \text{ between } x \text{ and } c \end{aligned}$$

Example 1: Finding a Taylor Series

Find the Taylor series for $f(x) = \cos x$, expanded about $x = \pi$. Then determine the interval of convergence.

Strategy. First, we the find some derivatives of $f(x) = \cos x$, and then we evaluate the derivatives at $x = \pi$. We can then place these derivatives in for the terms of the polynomial expansion.

We will use the ratio test in order to find the interval of convergence.

Solution.
$$\begin{array}{ll} f(x) = \cos x & f(\pi) = -1 \\ f'(x) = -\sin x & f'(\pi) = 0 \\ f''(x) = -\cos x & f''(\pi) = 1 \\ f'''(x) = \sin x & f'''(\pi) = 0 \\ f^{(4)}(x) = \cos x & f^{(4)}(\pi) = -1 \end{array}$$

We recognize that these derivatives have a pattern. All odd derivatives are zero, and the even ones alternate between $+1$ and -1. The Taylor series is

$$\sum_{k=0}^{\infty} \frac{f^{(k)}(\pi)}{k!}(x-\pi)^k$$

$$= f(\pi) + f'(\pi)(x-\pi) + \frac{f''(\pi)}{2!}(x-\pi)^2 + \frac{f'''(\pi)}{3!}(x-\pi)^3 + \frac{f^{(4)}(\pi)}{4!}(x-\pi)^4 + \ldots$$

$$= -1 + 0 + \frac{(x-\pi)^2}{2} + 0 - \frac{(x-\pi)^4}{4!} + 0 + \ldots$$

$$= -1 + \frac{(x-\pi)^2}{2} - \frac{(x-\pi)^4}{4!} + \frac{(x-\pi)^6}{6!} - \frac{(x-\pi)^8}{8!} + \ldots$$

$$= \sum_{k=0}^{\infty} \frac{(-1)^{k+1}(x-\pi)^{2k}}{(2k)!}$$

To find the interval of convergence we use the ratio test:

$$\lim_{k \to \infty} \left| \frac{a_{k+1}}{a_k} \right| = \lim_{k \to \infty} \left| \frac{(-1)^{k+2}(x-\pi)^{2(k+1)}}{(2(k+1))!} \cdot \frac{(2k)!}{(-1)^{k+1}(x-\pi)^{2k}} \right|$$

$$= \lim_{k \to \infty} \left| \frac{(x-\pi)^{2k+2}}{(2k+2)!} \cdot \frac{(2k)!}{(x-\pi)^{2k}} \right|$$

$$= \lim_{k \to \infty} \left| \frac{(x-\pi)^2}{(2k+2)(2k+1)} \right|$$

$$= 0$$

We know that $0 < 1$ for all x. Thus, the Taylor series converges for all x, and the interval of convergence is $(-\infty, \infty)$.

Summary. The Taylor Polynomial expansion can be easily remembered, because each term has the k-th derivative, $(x-c)^k$, and $k!$. In other words, all the *numbers* are the same beginning with $k = 0$. Also, remember that 0! is defined to equal 1.

Example 2: Estimating the Error in a
Taylor Polynomial Approximation

---

a) Use the Taylor Polynomial of degree 4 for $f(x) = \cos x$, expanded about $x = \pi$, to approximate $\cos 3$.

b) Estimate the error in the approximation of $\cos 3$.

c) Estimate the number of terms needed in the Taylor polynomial to guarantee an accuracy of 10^{-10}.

Strategy. We are going to use the expansion from example 1, which is an expansion for $\cos x$. In this problem we let $x = 3$, and we find the Taylor polynomial of degree 4 to approximate $\cos 3$.

Solution. a) Using the Taylor expansion from the previous expansion, we use P_4, the Taylor polynomial of degree 4.

$$P_4(x) = -1 + \frac{(x-\pi)^2}{2} - \frac{(x-\pi)^4}{4!}$$

$$\cos 3 \approx P_4(3) = -1 + \frac{(3-\pi)^2}{2} - \frac{(3-\pi)^4}{4!} = -0.9899925078$$

b) The remainder $R_n(x) = \left|\frac{f^{n+1}(z)}{(n+1)!}(x-c)^{n+1}\right|$ for $n = 4$, $x = 3$ and $c = \pi$ is given by

$$R_4(3) = \left|\frac{f^5(z)}{(5)!}(3-\pi)^5\right| \text{ for some } z \text{ between 3 and } \pi$$

$$R_4(3) = \left|\frac{-\sin z}{(5)!}(3-\pi)^5\right|$$

Note that $\sin z$ is always between -1 and 1, which gives us our bound on $f^{n+1}(z)$.

$$\implies |R_4(3)| \leq \left|\frac{(3-\pi)^5}{(5)!}\right| = (4.74266401) \cdot 10^{-7}$$

Thus the error in the approximation is less than $(4.8) \cdot 10^{-7}$.

c) In order to get an accuracy of 10^{-10}, we need the remainder to be smaller than 10^{-10}. All the derivatives of $f(x) = \cos x$ at $x = \pi$, are $\pm \sin x$, or $\pm \cos x$, thus $|f^{n+1}(z)| \leq 1$ for all z. Thus

$$|R_n(x)| = \left|\frac{f^{n+1}(z)}{(n+1)!}(3-\pi)^{n+1}\right| \leq \left|\frac{1 \cdot (3-\pi)^{n+1}}{(n+1)!}\right|$$

Since we need the remainder to be less than 10^{-10}, we solve

$$\left|\frac{(3-\pi)^{n+1}}{(n+1)!}\right| \leq 10^{-10}$$

Putting a bound on $|(3-\pi)^{n+1}|$ we have $|3-\pi| = 0.142... \leq 0.2$, and $|(3-\pi)^{n+1}| \leq |(0.2)^{n+1}|$ for all n. Thus

$$\left|\frac{(0.2)^{n+1}}{(n+1)!}\right| \leq 10^{-10}$$

(take the reciprocal of both sides which reverses the inequality)

$$10^{10} \leq \frac{(n+1)!}{0.2^{n+1}}$$

We need to find a value for n which satisfies this inequality. We find this value of n by trial and error.

If $n = 5$, $\frac{(n+1)!}{0.2^{n+1}} = 11,250,000$ is not large enough,

$n = 6$, $\frac{7!}{0.2^7} = 393,750,000$ is not large enough either.

However, $n = 7$, $\frac{8!}{0.2^8} = (1.575) \cdot 10^{10}$. Thus $n = 7$ guarantees an accuracy of 10^{-10}.

Summary. Once we found an expansion for $\cos x$, all we had to do was substitute 3 for x throughout the expansion. Basically, that was the easiest part.

To approximate the error, we used the remainder. Usually the hardest part about approximating an error is finding an upper bound for the derivative. However, when using either the sine or the cosine function, this becomes easy since both of these functions are bounded above by 1 and bounded below by -1.

To obtain an accuracy of 10^{-10}, (which is incredibly accurate), we used logic, and trial and error. We did not need to go higher than $n = 7$. This means that if we add up the first 7 terms of the Taylor polynomial we can approximate $\cos 3$ with an accuracy of 10^{-10}.

Example 3: Finding New Taylor Series from Old

Find the MacLaurin series for $f(x) = x \cdot \sin\left(x^2\right)$.

Strategy. We could find the derivatives of $f(x)$ directly, but that would give us a very involved calculation. It is faster to use the MacLaurin series for $g(y) = \sin y$.

$$\sin y = \sum_{k=0}^{\infty} \frac{(-1)^k y^{2k+1}}{(2k+1)!} = y - \frac{y^3}{3!} + \frac{y^5}{5!} - \frac{y^7}{7!} + \ldots$$

Solution. Replace y with x^2.

$$\sin\left(x^2\right) = \sum_{k=0}^{\infty} \frac{(-1)^k (x^2)^{2k+1}}{(2k+1)!} = x^2 - \frac{(x^2)^3}{3!} + \frac{(x^2)^5}{5!} - \frac{(x^2)^7}{7!} + \ldots$$

$$= \sum_{k=0}^{\infty} \frac{(-1)^k x^{4k+2}}{(2k+1)!} = x^2 - \frac{x^6}{3!} + \frac{x^{10}}{5!} - \frac{x^{14}}{7!} + \ldots$$

And

$$x \sin\left(x^2\right) = \sum_{k=0}^{\infty} \frac{(-1)^k x \cdot x^{4k+2}}{(2k+1)!} = x \cdot x^2 - x \cdot \frac{x^6}{3!} + x \cdot \frac{x^{10}}{5!} - x \cdot \frac{x^{14}}{7!} + \ldots$$

$$= \sum_{k=0}^{\infty} \frac{(-1)^k x^{4k+3}}{(2k+1)!} = x^3 - \frac{x^7}{3!} + \frac{x^{11}}{5!} - \frac{x^{15}}{7!} + \ldots$$

Summary. If you can use the Taylor series of a given function to find the Taylor series of another function, then this will be a lot easier than finding the Taylor series from scratch.

7.8. Applications of Taylor Series

Key Topics.
- Using Taylor Series to Evaluate Limits
- Using Taylor Series to Approximate Integrals

Worked Examples.
- Using a Taylor Series to Evaluate a Limit
- Using a Taylor Series to Approximate a Definite Integral

Overview

In this section, we see several applications of Taylor polynomials. In particular, we will evaluate limits and approximate integrals. As you know, working with polynomials is usually not very hard. Using polynomial expansions for transcendental functions will make us appreciate these expansions.

Using Taylor Series to Evaluate Limits. The Taylor or MacLaurin expansion can be used to find the value of difficult limits. This will be shown in one of the following examples.

Using Taylor Series to Approximate Integrals. Integrating polynomials is quite simple. It is, therefore, a great idea to use the polynomial expansion to approximate a definite integral. We can integrate a convergent series by integrating it term by term.

Example 1: Using a Taylor Series to Evaluate a Limit

Use a Taylor series to evaluate the limit $\lim\limits_{x \to 0} \dfrac{\sin x}{x}$.

Strategy. This limit can be evaluated with L'Hôpital's Rule. However, we will use a Taylor series to show how Taylor series can be used to find limits.

Solution. The Taylor series for $f(x) = \sin x$ is given by

$$\sin x = x - \frac{x^3}{3!} + \frac{x^5}{5!} - \frac{x^7}{7!} + \frac{x^9}{9!} - \ldots$$

Thus

$$\frac{\sin x}{x} = \frac{1}{x}\left(x - \frac{x^3}{3!} + \frac{x^5}{5!} - \frac{x^7}{7!} + \frac{x^9}{9!} - \ldots\right)$$

$$= 1 - \frac{x^2}{3!} + \frac{x^4}{5!} - \frac{x^6}{7!} + \frac{x^8}{9!} - \ldots$$

$$\lim_{x \to 0} \frac{\sin x}{x} = \lim_{x \to 0}\left(1 - \frac{x^2}{3!} + \frac{x^4}{5!} - \frac{x^6}{7!} + \frac{x^8}{9!} - \ldots\right)$$

$$= 1 - 0 + 0 - 0 + 0 - \ldots$$

$$= 1$$

Summary. Using the polynomial expansion does make it easier - just be careful and go slowly so that you do not mess up any of the terms. As you see, there is a lot of writing involved.

Example 2: Using a Taylor Series to Approximate a Definite Integral

Use the first three non-zero terms of the Taylor series for $y = \sin\left(x^2\right)$ to approximate the definite integral

$$\int_0^1 \sin\left(x^2\right)\, dx$$

Strategy. Since we can integrate Taylor series term by term, we can simply take the first three non-zero terms of the Taylor expansion for $\sin(x^2)$ and integrate these terms. We do not have to find the Taylor polynomial of $y = \sin\left(x^2\right)$ from scratch, we can find the terms for this expansion by substituting x^2 for x in the expansion of $\sin x$.

Solution. The Taylor series for $f(x) = \sin x$ is given by

$$\sin x = x - \frac{x^3}{3!} + \frac{x^5}{5!} - \frac{x^7}{7!} + \frac{x^9}{9!} - \cdots$$

Thus the Taylor series for $f(x) = \sin\left(x^2\right)$ is given by

$$\begin{aligned}
\sin\left(x^2\right) &= \left(x^2\right) - \frac{\left(x^2\right)^3}{3!} + \frac{\left(x^2\right)^5}{5!} - \frac{\left(x^2\right)^7}{7!} + \frac{\left(x^2\right)^9}{9!} - \cdots \\
&= x^2 - \frac{x^6}{3!} + \frac{x^{10}}{5!} - \frac{x^{14}}{7!} + \frac{x^{18}}{9!} - \cdots
\end{aligned}$$

Therefore

$$\begin{aligned}
\int_0^1 \sin\left(x^2\right)\, dx &\approx \int_0^1 \left(x^2 - \frac{x^6}{3!} + \frac{x^{10}}{5!}\right) dx \\
&= \left. \frac{x^3}{3} - \frac{x^7}{7 \cdot 3!} + \frac{x^{11}}{11 \cdot 5!} \right|_0^1 \\
&= \frac{1}{3} - \frac{1}{42} + \frac{1}{1320} \\
&= \frac{2867}{9240} \\
&= .3102813853
\end{aligned}$$

Summary. This approximation is based on only three terms of the polynomial expansion. Integrate $\int_0^1 \sin(x^2)\, dx$ on your calculator and see how amazingly close our estimate is.

7.9. Fourier Series

Key Topics.
- Fourier Series

Worked Examples.
- Finding a Fourier Series
- A Fourier Series for a Function of Period Other Than 2π

Overview

Taylor series expansions (power series) work extremely well as long as we stay within the interval of convergence. For values outside that interval, the series is not usable as a way to approximate the function. For periodic functions we can use another type of series called a Fourier series.

Fourier Series. For any series with period 2π, the Fourier series of a function is of the form:
$$\frac{a_0}{2} + \sum_{k=1}^{\infty} (a_k \cos(kx) + b_k \sin(kx))$$

The coefficients a_k and b_k of the Fourier series can be found as follows:

$$a_0 = \frac{1}{\pi} \int_{-\pi}^{\pi} f(x)\,dx$$

$$a_k = \frac{1}{\pi} \int_{-\pi}^{\pi} f(x) \cos(kx)\,dx$$

$$b_k = \frac{1}{\pi} \int_{-\pi}^{\pi} f(x) \sin(kx)\,dx$$

for $k = 1, 2, 3, \ldots$.

More generally, if f is periodic of period $T = 2l$, then the Fourier series is

$$\frac{a_0}{2} + \sum_{k=1}^{\infty} \left(a_k \cos\left(\frac{k\pi x}{l}\right) + b_k \sin\left(\frac{k\pi x}{l}\right) \right)$$

and

$$a_0 = \frac{1}{l} \int_{-l}^{l} f(x)\,dx$$

$$a_k = \frac{1}{l} \int_{-l}^{l} f(x) \cos\left(\frac{k\pi x}{l}\right) dx$$

$$b_k = \frac{1}{l} \int_{-l}^{l} f(x) \sin\left(\frac{k\pi x}{l}\right) dx$$

for $k = 1, 2, 3, \ldots$.

Example 1: Finding a Fourier Series Expansion

Find a Fourier series expansion for the function

$$f(x) = \begin{cases} 1 \text{ if } -\pi < x \leq 0 \\ 2 \text{ if } 0 < x \leq \pi \\ 1 \text{ if } \pi < x \leq 2\pi \\ 2 \text{ if } 2\pi < x \leq 3\pi \\ \text{etc.} \end{cases}$$

Strategy. This function has period 2π. Therefore, we will use a Fourier series of the form

$$\frac{a_0}{2} + \sum_{k=1}^{\infty} (a_k \cos(kx) + b_k \sin(kx))$$

We will find the coefficients as previously stated.

Solution.

$$a_0 = \frac{1}{\pi} \int_{-\pi}^{\pi} f(x)dx = \frac{1}{\pi} \left(\int_{-\pi}^{0} 1 dx + \int_{0}^{\pi} 2 dx \right) = \frac{1}{\pi} (\pi + 2\pi) = 3$$

And for $k = 1, 2, 3, \ldots$

$$\begin{aligned}
a_k &= \frac{1}{\pi} \int_{-\pi}^{\pi} f(x) \cos(kx) dx \\
&= \frac{1}{\pi} \left(\int_{-\pi}^{0} 1 \cos(kx) dx + \int_{0}^{\pi} 2 \cos(kx) dx \right) \\
&= \frac{1}{\pi} \left(\frac{1}{k} \sin(kx) \Big|_{-\pi}^{0} + \frac{2}{k} \sin(kx) \Big|_{0}^{\pi} \right) \\
&= \frac{1}{\pi} \left(\frac{1}{k} \sin(0) - \frac{1}{k} \sin(-k\pi) + \frac{2}{k} \sin(k\pi) - \frac{2}{k} \sin(0) \right) \\
&= 0
\end{aligned}$$

$$\begin{aligned}
b_k &= \frac{1}{\pi} \int_{-\pi}^{\pi} f(x) \sin(kx) dx \\
&= \frac{1}{\pi} \left(\int_{-\pi}^{0} 1 \sin(kx) dx + \int_{0}^{\pi} 2 \sin(kx) dx \right) \\
&= \frac{1}{\pi} \left(\frac{-1}{k} \cos(kx) \Big|_{-\pi}^{0} + \frac{-2}{k} \cos(kx) \Big|_{0}^{\pi} \right) \\
&= \frac{1}{\pi} \left(\frac{-1}{k} \cos(0) - \frac{-1}{k} \cos(-k\pi) + \frac{-2}{k} \cos(k\pi) - \frac{-2}{k} \cos(0) \right)
\end{aligned}$$

$$= \frac{1}{\pi}\left(\frac{-1}{k} + \frac{1}{k}\cos(-k\pi) - \frac{2}{k}\cos(k\pi) + \frac{2}{k}\right)$$

$$= \frac{1}{\pi}\left(\frac{1}{k} + \frac{1}{k}\cos(-k\pi) - \frac{2}{k}\cos(k\pi)\right)$$

Note that cosine is an even function, and, thus $\cos(-k\pi) = \cos(k\pi)$, so that

$$b_k = \frac{1}{\pi}\left(\frac{1}{k} - \frac{1}{k}\cos(k\pi)\right)$$

The Fourier series is

$$\frac{a_0}{2} + \sum_{k=1}^{\infty}(a_k \cos(kx) + b_k \sin(kx))$$

$$= \frac{3}{2} + \sum_{k=1}^{\infty}\frac{1}{\pi}\left(\frac{1}{k} - \frac{1}{k}\cos(k\pi)\right)\sin(kx)$$

Summary. You have probably noted that this is very tedious. However, the individual steps are not that difficult. You must just go very slowly to make sure you don't make any mistakes.

Example 2: A Fourier Series for a Function of Period Other Than 2π

Find a Fourier series expansion for the function

$$f(x) = \begin{cases} 1 \text{ if } -1 < x \leq 0 \\ 2 \text{ if } 0 < x \leq 1 \\ 1 \text{ if } 1 < x \leq 2 \\ 2 \text{ if } 2 < x \leq 3 \\ \text{etc.} \end{cases}$$

Strategy. The period of this function is 2. The general formula which we have for Fourier series has period $T = 2l$. Since the period of this function is 2, we have $T = 2, l = 1$.

Solution. The form of the solution is: $\dfrac{a_0}{2} + \sum_{k=1}^{\infty} \left(a_k \cos\left(\dfrac{k\pi x}{l}\right) + b_k \sin\left(\dfrac{k\pi x}{l}\right) \right)$.

$$a_0 = \frac{1}{1}\int_{-1}^{1} f(x)dx = \int_{-1}^{0} 1\, dx + \int_{0}^{1} 2\, dx = (1+2) = 3$$

And for $k = 1, 2, 3, \ldots$

$$\begin{aligned}
a_k &= \frac{1}{1}\int_{-1}^{1} f(x)\cos\left(\frac{k\pi x}{1}\right) dx \\
&= \int_{-1}^{0} 1\cos(k\pi x)dx + \int_{0}^{1} 2\cos(k\pi x)dx \\
&= \frac{1}{k\pi}\sin(k\pi x)\Big|_{-1}^{0} + \frac{2}{k\pi}\sin(k\pi x)\Big|_{0}^{1} \\
&= \frac{1}{k\pi}\sin(0) - \frac{1}{k\pi}\sin(-k\pi) + \frac{2}{k\pi}\sin(k\pi) - \frac{2}{k\pi}\sin(0) \\
&= 0
\end{aligned}$$

$$\begin{aligned}
b_k &= \frac{1}{1}\int_{-1}^{1} f(x)\sin\left(\frac{k\pi x}{1}\right) dx \\
&= \int_{-1}^{0} 1\sin(k\pi x)dx + \int_{0}^{1} 2\sin(k\pi x)dx \\
&= \frac{-1}{k\pi}\cos(k\pi x)\Big|_{-1}^{0} + \frac{-2}{k\pi}\cos(k\pi x)\Big|_{0}^{1} \\
&= \frac{-1}{k\pi}\cos(0) - \frac{-1}{k\pi}\cos(-k\pi) + \frac{-2}{k\pi}\cos(k\pi) - \frac{-2}{k\pi}\cos(0) \\
&= \frac{-1}{k\pi} + \frac{1}{k\pi}\cos(-k\pi) - \frac{2}{k\pi}\cos(k\pi) + \frac{2}{k\pi} \\
&= \frac{1}{k\pi} + \frac{1}{k\pi}\cos(-k\pi) - \frac{2}{k\pi}\cos(k\pi)
\end{aligned}$$

Note again that cosine is an even function, $\cos(-k\pi) = \cos(k\pi)$, and therefore
$$b_k = \frac{1}{k\pi} - \frac{1}{k\pi}\cos(k\pi)$$
The Fourier series is
$$\frac{a_0}{2} + \sum_{k=1}^{\infty}\left(a_k \cos\left(\frac{k\pi x}{l}\right) + b_k \sin\left(\frac{k\pi x}{l}\right)\right)$$
$$= \frac{3}{2} + \sum_{k=1}^{\infty}\left(\frac{1}{k\pi} - \frac{1}{k\pi}\cos(k\pi)\right)\sin(k\pi x)$$

Summary. Note that this looks very similar to the Fourier series in example 1. The only difference is that we have $k\pi$ in this example wherever we had just π in the previous example. The reason is that in the first example the period is 2π and in this example the period is 2.

7.10. Using Series to Solve Differential Equations

Key Topics.
- Using Series to Solve Differential Equations

Worked Examples.
- Using a Taylor Series to Solve a Differential Equation

Overview

In the previous chapter, we solved differential equations with constant coefficients. In this section we will solve differential equations with variable coefficients. As you probably have guessed, we will use series to do this.

Using Series to Solve Differential Equations. We start with a typical power series such as $y = \sum_{n=0}^{\infty} a_n x^n$. We will substitute this series into the equation and then use the resulting equations to find the coefficients.

Example 1: Using a Taylor Series to Solve a Differential Equation

Use power series to solve the differential equation $y'' = y$, $y'(0) = 1$, and $y(0) = 2$.

Strategy. We assume that we can write the solution as a power series. We can find the first and second derivatives of the power series and substitute the series we found for y'' and y in the equation. Then we will use the equation to determine the needed coefficients.

Solution. First we write y as a power series:

$$y = \sum_{n=0}^{\infty} a_n x^n = a_0 + a_1 x + a_2 x^2 + a_3 x^3 + ...$$

$a_0 = 2$, because $y(0) = 2$ and the first term is the only term without an x. Therefore, it is the only term that doesn't become 0 and drop out. Then

$$y' = \sum_{n=1}^{\infty} n a_n x^{n-1}$$

$a_1 = 1$, since $y'(0) = 1$, and, as before, only the first term does not have an x and doesn't drop out.

$$y'' = \sum_{n=2}^{\infty} n(n-1) a_n x^{n-2}$$

(rewrite as a sum with powers x^n)

$$= \sum_{n=0}^{\infty} (n+2)(n+1) a_{n+2} x^n$$

Note that this was accomplished by subtracting 2 from the index, which we then compensate for by adding 2 to every expression containing n. Write out several terms in each of the series and you will understand what was just explained. You will get two identical series.

Now, substituting the series into the given equation, $y'' = y$, we get

$$\underbrace{\sum_{n=0}^{\infty}(n+2)(n+1)a_{n+2}x^n}_{y''} = \underbrace{\sum_{n=0}^{\infty} a_n x^n}_{y}$$

and therefore, equating the coefficients of x^n:

$$(n+2)(n+1)a_{n+2} = a_n$$

$$a_{n+2} = \frac{a_n}{(n+2)(n+1)}$$

This defines a sequence recursively with $a_0 = 2$. Remember that above we determined that $a_0 = 2$ from $y(0) = 2$.

7.10. USING SERIES TO SOLVE DIFFERENTIAL EQUATIONS

Now when n is even, we have:

$n = 0$: $\quad a_2 = \dfrac{a_0}{2} = \dfrac{2}{2}$

$n = 2$: $\quad a_4 = \dfrac{a_2}{4 \cdot 3} = \dfrac{2}{4 \cdot 3 \cdot 2}$

$n = 4$: $\quad a_6 = \dfrac{a_4}{6 \cdot 5} = \dfrac{2}{6 \cdot 5 \cdot 4 \cdot 3 \cdot 2}$

$n = 6$: $\quad a_8 = \dfrac{a_6}{8 \cdot 7} = \dfrac{2}{8 \cdot 7 \cdot 6 \cdot 5 \cdot 4 \cdot 3 \cdot 2}$

From the pattern, you can see that to get the terms with even subscripts we have:

$$a_{2n} = \frac{2}{(2n)!}$$

When n is odd, we have $a_1 = 1$. Above we determined that $a_1 = 1$ from $y'(0) = 1$. Working as above:

$n = 1$: $\quad a_3 = \dfrac{a_1}{3 \cdot 2} = \dfrac{1}{3 \cdot 2}$

$n = 3$: $\quad a_5 = \dfrac{a_3}{5 \cdot 4} \cdot \dfrac{1}{5 \cdot 4 \cdot 3 \cdot 2}$

$n = 5$: $\quad a_7 = \dfrac{a_5}{7 \cdot 6} = \dfrac{1}{7 \cdot 6 \cdot 5 \cdot 4 \cdot 3 \cdot 2}$

From the pattern, you can see that to get the terms with odd subscripts we have:

$$a_{2n+1} = \frac{1}{(2n+1)!}$$

We now have a power series solution:

$$y = \sum_{n=0}^{\infty} \frac{2}{(2n)!} x^{2n} + \sum_{n=0}^{\infty} \frac{1}{(2n+1)!} x^{2n+1}$$

Summary. Once again, we have something which is very tedious, but not necessarily very difficult. If you keep track of what you are doing, and go very slowly, then you should be fine. Just carefully figure out the series for the functions in the equation.

Notes